THE RESTLESS
KINGDOM

THE GEOLOGICAL TIME SCALE

Era	Period (Epoch)	duration (millions of years)	years from today (millions of years)	Animal life
CENOZOIC				
	Quaternary			
	Recent	.01	.01	Rise of Civilization
	Pleistocene	1	1	Genus *Homo*
	Tertiary			
	Pliocene	12	13	
	Miocene	12	25	
	Oligocene	11	36	Age of Mammals
	Eocene	22	58	
	Paleocene	5	63	
MESOZOIC				
	Cretaceous	72	135	Last of the Dinosaurs
	Jurassic	46	181	Age of Dinosaurs, First Birds
	Triassic	48	230	First Dinosaurs, First Mammals
PALEOZOIC				
	Permian	50	280	Expansion of Reptiles, decline of Amphibians
	Carboniferous	65	345	Age of Amphibians, Abundant Insects
	Devonian	60	405	Age of Fishes, First Insects and Amphibians
	Silurian	20	425	First invasion of Land by Arthropods
	Ordovician	75	500	First Vertebrates
	Cambrian	100	600	Abundant Marine Invertebrates
PROTEROZOIC		1000	1600	The First Invertebrates
ARCHEOZOIC		2000	3600	Earliest Life Forms

THE RESTLESS KINGDOM

An Exploration of
Animal Movement

JOHN COOKE

Facts On File
New York • Oxford

Facts On File, Inc.
460 Park Avenue South
New York NY 10016
USA

Library of Congress Cataloging-in-Publication Data
Cooke, John A. L.
   The restless kingdom: an exploration of animal movement / John Cooke.
      p.   cm.
ISBN 0-8160-1205-9
1. Animal locomotion.   I. Title.
QP301.C589   1991
591.1'852–dc20   90-34488

Facts On File books are available at special discounts when purchased in bulk quantities for businesses, associations, institutions or sales promotions. Please contact the Special Sales Department of our New York office at 212/683-2244 (dial 800/322-8755 except in NY, AK or HI).

Composition by TSI Graphics
Manufactured by Mandarin Offset
Printed in Hong Kong

10 9 8 7 6 5 4 3 2 1

This book is printed on acid-free paper.

TABLE OF CONTENTS

The cheetah is the most highly adapted of the great cats for high speeds and maneuverability. Its most characteristic features are large chest capacity, long legs, light build and extreme flexibility of the spine to provide a long stride.

PREFACE

This book is a journey of exploration through the animal kingdom examining the ways in which animals move. As such, it is broadly scientific, but it is not a textbook and it is not comprehensive. It is meant to be read for pleasure as well as enlightenment, and consequently the approach throughout has been selective. The new, the remarkable and the improbable — the things that I personally have found exciting — are accorded more space than the old, the familiar and the mundane. As a general rule, if a topic could be talked about enthusiastically at a social gathering without generating widespread suspicions of paranoia, idiocy or inebriation, it was a candidate for inclusion.

The viewpoint throughout the book is biological and goes no further into mechanics than is necessary for a broad understanding of the topics covered. Although the use of some technical language is unavoidable in a work of this kind, no special knowledge is assumed and all unfamiliar terms are, it is hoped, adequately defined.

This book reports no original discoveries, but is a winnowing of research and reviews that have caught my eye and my fancy. Some material has been omitted intentionally but I hope little that would have merited inclusion has been overlooked.

It is customary at this point to thank those without whose help and patience volumes of this kind could not be written. Foremost in this category must be my wife, Joyce, whose warm support and encouragement have carried me equally through periods of disenchantment and obsession. I also owe her a special thanks for the gift of a powerful computer. This not only greatly facilitated the mechanics of writing, but actively contributed to the quality of the finished text. On a number of painful occasions the machine, doubtless improperly instructed, steadfastly refused to divulge the secrets previously committed to its memory. The rewritten text always benefited from the resulting extra labor.

Thanks are also due to the friends and strangers alike who, on asking about the book, were patient or polite enough to listen to my current enthusiasm, so bolstering my confidence and will to continue. Likewise the library staffs at the American Museum of Natural History, New York, and the University of California, Berkeley, for their cheerful help and advice. Jean and Theo Roland-Entwistle, Bernard Furnival, Taylor Kingsley, Professor David Nichols and Dr. John Edwards read drafts of the text and eliminated numerous errors of fact, logic and grammar. The mistakes and omissions that remain are wholly my own.

Special thanks are also due to Tony Tilford for the use of his Presto high-speed flash system and numerous other kindnesses, and for the generous use of many of his photographs to illustrate this volume.

J.A.L.C.

INTRODUCTION

Fascination with animal movement is as old as mankind, but the subject has always remained one of nature's great mysteries. Even the legendary wisdom of Solomon himself was confounded by "the way of an eagle in the air; the way of a serpent upon a rock" (Proverbs 30:19). It is the power of movement, more than anything else, that clearly sets animals apart from plants — movement of infinite variety and beguiling beauty that represents a remarkable compromise between size, structure and mechanical perfection. It is often tempting to draw comparisons between man-made devices and the natural world, but the conclusions will not always be valid. Engineers may be able to propel us more rapidly than can nature unaided, but when the total system is considered — speed, agility and economy of effort — the outcome of millions of years of natural selection still wins by a substantial margin.

Locomotion by machine relies heavily on the continuously rotating wheel, a form of movement almost never found in nature. Natural movement depends on mechanisms that are, as yet, too complex for engineers to imitate in detail and that rely on highly coordinated and sophisticated sensory and control systems.

The scientific study of locomotion got off to a slow start. While it is true that Leonardo da Vinci (1452–1519) was fascinated by flight and other forms of movement, his ideas had no impact, for they remained unknown until his notebooks were edited and published in 1930. And although *De motu animalium ex principio statico* by Italian physicist and astronomer Giovanni Borelli (1608–1679) was published in two volumes in 1680–81, the first significant contribution to the technical literature was not made until 1873, when the French physiologist Jules Marey (1830–1904) published his book *Machine animale*. Since then, the pace has quickened considerably, and scarcely a week now passes without the announcement of some fresh discovery.

For many years research into animal locomotion fell between the two stools of biology and mechanics. Those who went into biology often did so in order to escape the mathematical discipline of the "hard" sciences, and few biologists had any grounding in mechanical engineering. At the same time, few engineers felt inclined to carry their investigations into the enormously complex and unpredictable world of living organisms. Rigid models in a controlled laboratory setting provided more than enough problems. Today, this situation is changing. With the proliferation of interdisciplinary studies, there are now many biologists well versed in mathematics and mechanics, and not a few engineers with penetrating curiosity about the realities of the natural world.

In this book we shall be examining the diverse ways in which animals move, and the adaptations that have evolved to enable them to do so efficiently. It is a scientific book that deals with both biology and mechanics, and as such may involve concepts and ideas that are new or unfamiliar. Although the language of science is rooted in mathematics, this book is intended for the general reader and so the approach is qualitative rather than quantitative. For this reason all but the simplest mathematical notation has been excluded from the text, but it has not been possible to avoid the use of scientific language. However, a glossary of terms that may be unfamiliar appears at the end of the book.

All weights and measures are presented in units of the international scientific metric system. Hence distances are given in meters (1.09 yard) and kilometers (0.621 mile) and weights in kilograms (2.20 pounds) and tonnes (0.984 ton). The micron (μ) is a unit of length equal to 1/1000 millimeter. Temperatures are all given in degrees Celsius (°C) and frequencies, such as the rate at which a bird beats its wings, are given in hertz (Hz), the equivalent of cycles per second. It should be noted that the term "billion" is used in its American sense, meaning one thousand million, rather than its British sense of one million million.

In the first chapter we begin by examining some of the reasons why animals move and the constraints imposed by particular types of movement. Then a number of essential background topics are considered, including the structure and function of muscles, the nature of skeletal support and the significance of size. Finally, because this book is essentially about evolution, the major elements of natural selection are discussed.

The second chapter deals with background topics that are essentially mechanical rather than biological in nature. These include a survey of the laws of motion, the elements of fluid dynamics and the story of the discovery of the principles of aerodynamics.

These first two chapters should be viewed as an integral part of our story, for in large measure they deal with the common threads that unite themes pursued in later sections.

The arrangement of subsequent chapters reflects the broad course of evolutionary history, starting with life in water, progressing on to land and finally taking to the air.

In the course of these pages we shall travel the length and breadth of the animal kingdom, encountering groups with which even most zoologists would not claim familiarity. At the end of the book is an appendix that summarizes this sizable cast of characters, together with the salient features of their organization and morphology.

BACKGROUND BIOLOGY

INTRODUCTION

Painters and poets, philosphers and physiologists have all marveled at the way animals move and rejoiced in the grace and beauty of their doing so. Plants and animals differ from each other in numerous and fundamental ways, but nothing sets animals apart more conspicuously than their power of locomotion. All animals, at some stage in their life history, possess the ability to move from place to place under their own power, a capability conspicuously absent among plants.

For many animal species, an individual's very survival is dependent on its continuing mobility. Equally, there are others that have successfully adopted a totally sedentary existence for the greater part of their lives. For these, it is the precise location of their chosen home that is of primary importance, with the heavy responsibility of site selection resting with a diminutive mobile larval stage. Indeed, dispersal and the choice of a suitable location in which to settle appear to be the very reasons for the existence of such a stage in the life history of these animals. One might wonder what benefits a sedentary life can bestow. An obvious advantage is the marked reduction in energy needs, which makes possible the exploitation of habitats low in food resources. To some extent this is offset by increased vulnerability to attack, but most sessile creatures live either buried or within a secreted protective retreat. Although building such a retreat is a major energy investment, its growth, like that of the animal itself, is directly related to the available food supply.

For free-living animals, locomotion is necessary for a wide variety of purposes, each imposing its own special demands. The exact system of movement found in any particular species is an expression of the optimum balance between a multitude of conflicting demands. At all times, selection acts to eliminate waste and conserve only that which is necessary for survival.

The leopard normally hunts by night and sleeps camouflaged by day. It is built for short, powerful attacks.

Predators, for example, need be able to run no faster than is necessary for them to capture prey with a reasonable level of success. The intended victim, however, does not necessarily have to escape its nemesis through speed alone. A small, agile animal that waits until the last moment to swerve can escape a predator that is significantly fleeter of foot.

Migratory species must carry sufficient fuel to cover the required distance at a speed that provides the greatest economy. Evolution has determined the correct balance between the amount of energy needed to transport fuel and the rate of its consumption to maximize performance. Too much fuel for the journey is little better than not enough, while moving too quickly or too slowly will each result in failure to arrive. The point is that the unique requirements of each species will be reflected in the variety of locomotory mechanisms and components it possesses. The muscular and energy requirements of a conspicuous species that depends on speed and agility to escape predators will be very different from a close relative that relies on stealth and camouflage for its protection. Muscles that maintain posture are quite different in performance from those for rapid motion, contracting only slowly but with minimal power consumption.

But movement is not always a question of migration or predation. Courtship is often associated with movement, as is the resulting production and care of young. Elaborate displays between sexes are found in many animal groups, and are frequently followed by nest building, excavation and similar activities. All of these require the addition of specific kinds of movement to the animal's locomotory armory.

Sometimes the need for locomotion can be met by behavioral rather than mechanical solutions. Small terrestrial arthropods often face a daunting challenge when they need to disperse. Some, such as spiders, are able to employ strands of silk to make use of warm updrafts of air and be carried

Impala are lightly built and very agile. Their ability to change direction abruptly often enables them to escape swifter predators.

Swallows hunt insects on the wing and are fast, maneuverable fliers. Because their prey is seasonal, swallows undertake long migrations to maintain their food supply, and often travel in huge flocks, like these seen in Florida in early spring.

Most spiders disperse soon after hatching by ballooning — riding rising air currents on a silken thread. Money spiders like this one are small and can travel hundreds of miles even when fully grown.

hundreds of kilometers at almost no energy cost, a process known as ballooning. An exact marine counterpart to ballooning is found among certain tiny snails that drift on the ocean currents suspended on threads of secreted mucus. Others, such as pseudoscorpions, lacking the ability to balloon like spiders, depend on hitchhiking on flies and beetles — or phoresy, to give this method of dispersal its scientific name.

THE POWERHOUSE

Many microscopic organisms rely on protoplasmic streaming or on the use of cilia and flagella for propulsion. All other forms of animal locomotion depend ultimately on the contraction of muscle fibers. Although various kinds of muscle fiber are found in the animal kingdom, all appear to be based on a fundamentally similar molecular system of sliding protein filaments (Fig. 1). That such a system should prove so efficient, so adaptable and so widespread is indeed remarkable.

Muscles are built up of individual fibers, each of which is composed of numerous myofibrils, which in turn are composed of filaments made of the proteins actin and myosin. Each muscle fiber is either a single cell or a syncytium, an aggregation of fused cells. The myofibrils that make up a muscle fiber are composed of functional units called sarcomeres arranged end to end. These are the basic contractile elements of muscle.

Among vertebrates, three main types of muscle are recognized: skeletal (or striated), smooth (or visceral) and cardiac. We need not be concerned here with the characteristics of contraction and coordination that distinguish these different kinds, for we are interested only in skeletal muscle, the one type used in vertebrate locomotion.

Vertebrate skeletal or striated muscle (and also that used for locomotion by many invertebrates) is characterized by a distinct pattern of cross-striations when viewed under the light microscope. This pattern of striations is repeated roughly every two microns, and indicates where thick myosin filaments interdigitate very regularly with thin actin ones within each myofibril. Each repetition from Z-line to Z-line comprises a single sarcomere. Molecular cross-bridges from the thick myosin filaments attach to the thin actin ones, drawing the two sets of filaments into one another, like a person pulling in a rope hand over hand.

Contraction requires energy, and this is provided by the breakdown of ATP (adenosine triphosphate) to ADP (adenosine diphosphate) on the cross-bridges. This is part of a complicated cycle of biochemical energy processes that occur within living cells. In fact, there are two major types of

Pseudoscorpions cannot balloon. Because of their small size they are able to hold onto the legs of flies and other insects and so disperse by phoresy, a form of hitchhiking.

STRUCTURE OF MUSCLE

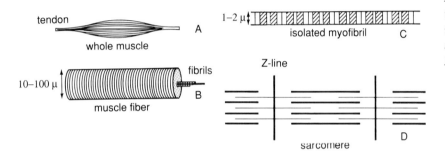

Fig. 1 The Structure of Muscle: (a) Whole muscle; (b) Single muscle fiber; (c) Isolated myofibril; (d) Single sarcomere within a myofibril, showing the relationship between actin filaments (thin) and myosin filaments (thick). (After Barrington.)

Small caterpillars often travel long distances on air currents by hanging from strands of silk. This Trinidadian species gains protection from predators by resembling the drifting seed of a forest tree.

metabolic pathways involved, aerobic and anaerobic. Most muscles require the presence of oxygen in order to function, and the rate at which this can be supplied is often the limiting factor in muscle performance. The centers of energy production in this aerobic metabolism are the mitochondria, small organelles scattered throughout cells, and particularly abundant in muscle.

Crabs and vertebrates (including humans) have both evolved the capacity for anaerobic metabolism. In this process, glucose is converted to lactic acid without the immediate need for oxygen, allowing very rapid release of energy. However, the resulting buildup of lactic acid, accumulating in the muscle tissues, cannot be tolerated indefinitely. Eventually oxygen will be required to convert the lactic acid back to glucose. This process is regarded as the repayment of an oxygen debt, and is the reason for extended deep breathing after heavy exercise.

Much of the ATP from which cells obtain their energy is derived from the breakdown of carbohydrates. However, carbohydrates and simple sugars are not the only fuel source. Both fats and proteins may also be used in this way through interlinked metabolic pathways. In fact, weight for weight, fat yields twice the energy of carbohydrate and is sometimes used as a concentrated source of power by migrating animals such as locusts.

Although the length of a myofibril, and hence its mechanical properties, can vary greatly, the length of an individual sarcomere is generally between 1.5 and 2.5 microns, although this range is sometimes exceeded among crustaceans. Because these long crustacean sarcomeres are composed of abnormally long actin and myosin filaments, possessing a greater number of cross-bridges than usual, their contractions are more powerful.

However, many short sarcomeres, end to end, can contract more rapidly because each has fewer cross-bridges to form, so such crustacean muscles are relatively slow-acting. In fact, the rate of shortening can vary enormous-

Insects such as this katydid are able to conserve energy by evolving a close resemblance to an inert model such as a dead leaf.

ly between different muscles. Some, like the limb muscles of mice and those of the raptorial pincers of mantis shrimp (*Squilla*), contract within milliseconds. Others, such as the walking muscles of tortoises and the muscles that help maintain posture in birds, can take several seconds to contract. Such slow contractions are far more energy efficient than fast ones.

Among vertebrates we shall encounter these different kinds of muscle performing specific tasks within a single animal. Thus fish use red aerobic muscle for regular swimming, but are able to carry substantial amounts of white anaerobic muscle as well for rapid escape maneuvers.

Mammals possess three types of fiber, two being red and well provided with mitochondria. These are the aerobic SO (slow oxidative) and the FOG (fast oxidative glycolytic), which can be both aerobic and anaerobic. The third type, FG (fast glycolytic) is anaerobic. All three types of fiber are found in varying proportions in the leg muscles of animals, and are used differentially. Thus few FG-type fibers are used in slow running, but many in fast running.

The force generated by each muscle fiber is proportional to the number of parallel cross-bridges in each sarcomere. Thus muscles with long filaments, such as those in the legs of grasshoppers, can exert more than twice the pull of normal vertebrate muscle. However, the grasshopper gains this advantage at the expense of speed of contraction. This may appear a contradiction, as jumping necessitates an extremely high acceleration on take-off. This apparent conflict of interest is solved for the grasshopper by the presence of a mechanism that stores the energy, generated relatively slowly by the contraction of the leg muscles, and releases it with explosive suddenness at the moment of jumping.

There is a limit to the rate at which cross-bridges can be formed, and the fastest contractions occur when the fibers are short and lightly loaded. The

Kangaroos use their tails as a third leg for slow, local travel. Their powerful hind legs contain massive tendons that store elastic energy during jumping.

fibrillar flight muscles of higher insects are of a special type and capable of very rapid operation. When activated under tension, fibrillar muscle oscillates at frequencies of up to 1000 Hz, a single nerve impulse serving to trigger a train of oscillations. Fibrillar muscles are able to function at these high frequencies partly because they undergo very small changes in length and partly because their contractions, once started, are repetitively triggered from within. This means that, unlike other muscles, their frequency of contraction is not limited by the rate at which nerve impulses can be transmitted.

Just as a muscle that shortens under tension does work, a muscle that is forcibly extended has work done on it. This occurs, for example, when a jumping kangaroo decelerates on landing. In this case negative work is being done, but metabolic energy is still being consumed and released as heat. This type of situation occurs frequently in animal locomotion, and has led to the evolution of a variety of natural elastic substances that can store the energy generated by doing negative work and release it moments later for the next propulsive cycle. Such energy storage systems are found in the tendons of mammals, the hinges of bivalve mollusks and in the legs and wings of insects. More are being identified all the time as researchers investigate the mechanics of ever more animal species.

The arrangement of fibers within a muscle shows a number of significant variations. For example, in some invertebrates such as nematodes, annelids and mollusks, the striated fibers are arranged in a helical pattern, some changes in length being effected by altering the pitch of the spiral. Some invertebrates, particularly those with hydrostatic skeletons that make substantial changes to their body shape, also use smooth muscle for locomotion. Vertebrates use smooth muscles for involuntary, nonlocomotive purposes.

In many arthropods particularly, the muscle fibers, instead of running parallel to each other, converge on a central tendon and are thus inclined to

The hind legs of grasshoppers contain powerful muscles for jumping, but they cannot contract rapidly enough to provide sufficient acceleration for takeoff. The energy of the muscles must be stored and then released suddenly when needed. Grasshoppers are generally good fliers once they have become airborne.

Moths, such as this yellow underwing (Noctua), have flight muscles that require a nerve impulse for each contraction. This limits the frequency of their wing beats to less than 100 Hz. Nevertheless, the noctuids — the family to which this species belongs — are strong and very maneuverable fliers.

one another. The angle of insertion changes as the fibers contract. Pennate muscles, as this arrangement of fibers is called, possess special properties. First, the shorter, more numerous fibers can impart much greater force than the parallel fibers of conventional muscle, albeit over a shorter distance. Also, whereas contraction of parallel fibers results in a significant swelling of the muscle, even though its total volume remians essentially constant, such swelling does not occur in pennate muscles. This makes them particularly suited to the rigid, armored bodies of arthropods, where the design of the limb joints also places a premium on strength rather than extensibility. Pennate muscles are found, for example, in the pincers of crabs and lobsters.

COORDINATION AND CONTROL

Even the simplest types of locomotion involve the contraction of numerous muscle fibers in a specific sequence and with precise timing. As we progress up the evolutionary ladder, patterns of locomotory activity become increasingly complex and variable. The control of such activity involves an extremely subtle interplay of nervous activity that integrates genetically preprogrammed neural pathways with input from myriad sense organs.

This is a field in which biologists still have only a very basic understanding of the principles involved, and as yet there is no animal in which the full story has been worked out. Moreover, there are clearly important differences between different groups of animals, and particularly between vertebrates and invertebrates.

Most vertebrate muscle fibers contract on an all-or-nothing basis. That is to say, all the fibers associated with a single neuron contract fully upon the arrival of a nerve impulse, and variations in the strength of contraction of a muscle depend on the number of neurons and hence on the number of fibers activated. Where very fine control is required, for example in the muscles of the eye, each fiber has its own neuron, but in the major limb muscles that make large, powerful movements, 200 or more fibers may be controlled through a single neuron.

Among invertebrates the situation is generally different. Here each fiber has one or more neurons, but the effect of a single nerve impulse is localized. Repeated impulses depolarize an increasing area of the fiber membrane, inducing more and more of the fiber to contract. Thus the extent of contraction in these muscles depends on the frequency with which impulses are received.

In normal invertebrate locomotion, such as that found in a burrowing worm, a regular cycle of muscle activity is maintained, usually derived from some form of metachronal rhythm progressing from segment to segment along the body. It used to be thought that coordination resulted from the completion of one muscle contraction stimulating the next in a chain of preordained reflexes. It is now recognized that the pattern is programmed into the central nervous system, and once initiated, runs independently. This is not to say that external influences can exert no impact.

A network of sensory cells feeds back into the system a detailed picture of how the limbs are working. There appears to exist a sort of template of anticipated patterns of input. So long as the actual pattern corresponds to this expected one, the system proceeds without modification. On the other hand, an unexpected change in direction or a slipping foot will immediately initiate appropriate compensatory action. Whatever the actual mechanism of control and coordination turns out to be, there is no question that it is enormously complicated and extremely sophisticated.

Ghost crabs (Ocypode) *are among the few crustaceans that have become successful on land. They are active predators on beaches and are capable of brief bursts of high-speed running, particularly to escape being eaten by birds. However, their respiratory system severely limits their endurance.*

In addition to normal locomotion, most animals possess special high-speed escape behaviors. These are short-lived bursts of activity that are called into being when an animal's survival is threatened. As such, they must subordinate all other activities and utilize the fastest pathways.

This usually involves the use of special nerve fibers and associated circuits to inhibit normal activity and sensory input temporarily.

SKELETONS

We are reasonably familiar with the design of our own bodies, and can readily appreciate the way in which movements are generated by muscles acting on a rigid internal skeleton. Such a system is common to all vertebrates, with a wide range of specific and relatively minor variations suited to particular life-styles. In contrast, invertebrates make use of several markedly different types of skeleton. These may be more or less rigid like the internal and external shells of many mollusks, or the plates that make up the test of echinoderms. Or they can be composed of rigid elements that articulate against one another like the armored external skeleton of arthropods. However, the first skeletons to appear on the evolutionary ladder were fluid-filled. These hydrostatic skeletons, as they are called, contain a fixed volume of liquid whose shape is altered by muscular contractions. Many invertebrates still depend for their movement on some variant of this type of skeletal system.

Perhaps the simplest hydrostatic skeleton is that found in sea anemones. The body cavity of anemones opens to the outside world through a simple pharynx at the base of the ring of tentacles. This opening serves as both mouth and anus. Clearly a system of this kind, with no special valves or seals, cannot generate much pressure, but anemones are able to maintain a slight positive internal water pressure by means of the siphonoglyph, a groove lined with cilia, whose synchronized beating wafts water into the

Each segment of an earthworm's body is a self-contained mechanical unit with its own hydrostatic skeleton and antagonistic muscles. This makes it particularly well suited for moving through the burrows it excavates in the soil.

body cavity. The slow, steady current of water enables the anemone not only to maintain its shape, but also to restore it after collapsing with the ebb tide. In some species the pumping action of the siphonoglyph is even sufficient to support active burrowing into sand and mud.

In more advanced hydrostatic skeletons, like those of the annelid worms, the fluid-filled body cavity, the coelom, is effectively isolated from the outside. In a cylindrical animal possessing both circular and longitudinal muscles, these will act antagonistically through the fluid in the coelom, which thus acts as a hydrostatic skeleton. Contraction of the circular muscles will elongate the body, while contraction of the longitudinal ones will shorten it. For more elaborate movements, it is often advantageous for the fluid of the hydrostatic skeleton to be separated into smaller functional units. In earthworms, for example, the coelom within each segment is effectively isolated from its neighbors by intersegmental septa, or partitions.

In leeches, annelid relatives of the earthworm, the coelom is greatly reduced and has lost its skeletal function. This role is taken over by botryoidal tissue, a specialized type of connective tissue that has invaded the coelomic spaces. By providing a very stiff alternative hydrostatic skeleton, botryoidal tissue is particularly well suited to the needs of the leeches' specialized form of locomotion.

A cellular hydrostatic skeleton is also found among phyla too simple to possess a coelom, such as flatworms and ribbon worms. The bodies of these creatures are filled with gelatinous cells called parenchyma, which function similarly to the botryoidal tissue of leeches. However, these acoelomate worms have a much simpler musculature and place fewer demands on their skeletal system.

In mollusks the skeleton of the foot is formed from the hemocoel, a blood-filled body cavity similar to that of arthropods, but which has become subdivided into numerous small interconnected spherical chambers. The principle of antagonistic pairs of muscles acting on some form of skeletal system is almost universal among the metazoa, or multicelled animals, and is usually an essential component of their locomotion.

In slugs and snails, such as the large tropical Achatina, *the hydrostatic skeleton is broken down into innumerable tiny blood-filled compartments scattered throughout the foot. Complex musculature allows some species to move parts of the foot independently, so enabling the animals to move with a variety of gaits.*

SIZE AND SCALING

No single factor influences locomotion more profoundly than size. Living organisms exist in an extraordinary size range that covers 21 orders of magnitude. This is to say that the largest, a blue whale weighing about 100 tonnes, is 10^{21} times more massive than the smallest microbe. In this section the implications of this disparity in size will be examined.

The earliest life-forms to appear were all small and single-celled. It is significant that even after 3.5 billion years of evolution (see Table 1, page ii), the complex chemistry of life can still function only in units a few microns in diameter. Increase in size and complexity is possible only if cells aggregate and specialize in function to form multicellular organisms. From the fossil record it is found repeatedly that there is a strong tendency toward increase in size, and it is not hard to think of possible reasons why this might be so. These will be considered in more detail below.

However, it is not sufficient for evolutionary progress that a successful animal design simply be uniformly magnified. Scaling up or down, as it is called, is much more complicated than simple change in size, and necessitates changes in proportion as well as magnitude. These changes result, in part, from the relationship between an object's surface area and its volume, a principle first recognized by Aristotle. In a sphere, volume is proportional to $(radius)^3$ and its surface area to $(radius)^2$. Hence volume (and therefore mass) increases as the cube of linear dimension but surface area increases only as the square, which means that surface area is proportional to $(volume)^{2/3}$ or $(volume)^{0.67}$.

For animals, the relationship between volume and surface area has many important mechanical and physiological consequences. Heat loss, gas exchange and the assimilation of food all take place across surfaces. The strength of a bone or a limb is related to its cross-sectional area, as is the power consumption of a muscle. The lift generated by a wing is related to its surface area, as is the drag resulting from skin friction on a swimming animal. Likewise, the effects of gravity, inertia and metabolic processes are related to volume and increase far more rapidly with increasing size.

When two measurements, for example limb length and diameter, vary in uniform proportion in individuals of different sizes, they are said to be geometrically similar, or isometric. As we have seen, however, in nature it is more usual for proportions to vary with size, in which case the measurements vary not isometrically but allometrically. This allometric relationship between the relative sizes of two parts can be expressed mathematically, and provides us with a valuable tool for comparing data. For example, if measurements of the surface area and mass in different species of vertebrate are plotted on logarithmic graph paper, the points are found to lie on a straight line with a slope of 0.63. This is another way of saying that the surface area of mammals is proportional to $(volume)^{0.63}$. Although this does not appear very different from the theoretical exponent 0.67 we saw above, it is nevertheless significant, as we shall soon discover.

Kleiber's Law

If we now look at the basal metabolic rate (measured by either oxygen consumption or heat production) we find that within a species it is proportional to $(mass)^{0.67}$ for individuals of different sizes. This fits with the traditional view that metabolism scales in proportion to surface area because both heat loss and gas transfer are surface phenomena. However, if different-size mammal species are compared, the relationship breaks down dramatically, with basal metabolic rate proportional not to $(mass)^{0.67}$ but to $(mass)^{0.75}$.

Flatworms like Polycelis *are among the least complex multicellular animals. Their simple skeleton, made of gelatinous cells, allows only rudimentary peristaltic movements. Many flatworms depend for their locomotion on a layer of beating cilia that cover the underside of the body.*

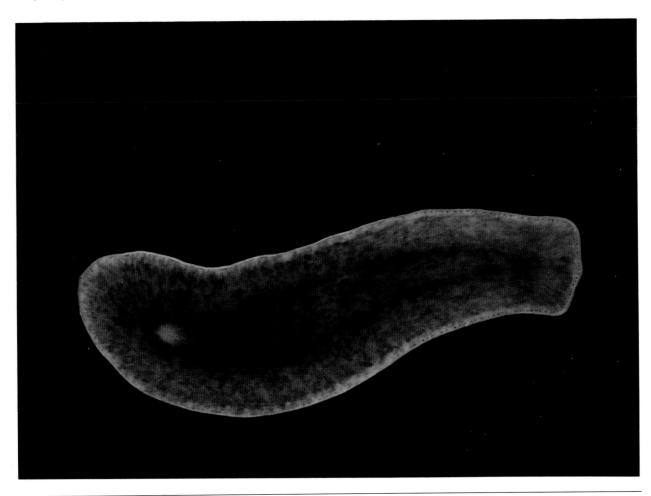

Known as Kleiber's law, after its discoverer, the American veterinarian Max Kleiber, this allometric relationship quantifies the observation that the pace of life slows down as animal-size increases.

The most dramatic illustration of Kleiber's law in action is to be found in mammals at the time of birth. A newborn human infant metabolizes (and so consumes oxygen) at a level as though it were simply an organ within the body of its mother. Within a day and a half of independent life, its metabolic rate doubles and reaches the level predicted by Kleiber's law appropriate for an organism of its actual size.

Elastic Similarity

Kleiber's law strongly suggests that something is amiss in the assumptions that we have been making about scaling and animal proportions. Predictions are found to accord more closely to reality if the figures are derived somewhat differently. This alternative relationship, which has only been recognized quite recently, is known as elastic similarity.

When a typical quadruped such as a dog, antelope or horse runs, its skeleton, and particularly its long bones, are subjected to very powerful forces, especially as it touches ground after all four feet have been airborne. Moreover, the larger the animal, proportionately greater are the forces to which it is subjected. This is reflected in the relative thickness of such bones, which increases allometrically with size. In fact it has been recognized empirically for a long time that the diameter of long bones is pro-

Dogs are descended from hunting ancestors that were strong runners. Selective breeding has given rise to many domestic varieties, some designed for speed and others for tenacity or aggression. The ancestral form represents a compromise between the competing virtues of strength, agility and endurance.

portional to (length)$^{3/2}$. From this it has been possible to deduce a lot about the life-style and relative athleticism of extinct fossil forms from their incomplete skeletal remains.

Under the rules of elastic similarity the body of an animal is regarded as being made up of numerous cylinders of differing diameters. For each element the *length × (diameter)²* is proportional to total body mass. From this it follows that the diameter of each segment is proportional to (mass)$^{3/8}$ and the length proportional to (mass)$^{1/4}$.

How does this relate to Kleiber's law? The pull of a muscle, and hence its oxygen consumption, is related not to body size but to its cross-sectional area. If we were to consider metabolic rate as being proportional to cross-sectional area, what figures would we get? Under the rules of elastic similarity, as we saw above, diameter is proportional to (mass)$^{3/8}$. Hence cross-sectional area must be proportional to (mass)$^{3/4}$. This, it will be recalled, is the empirical value that Kleiber observed.

The implication that metabolic rate is actually limited by cross-sectional area rather than by surface area is also borne out by measurements of peak metabolic performance (measured as oxygen consumption) in a wide range of mammals from pygmy mice to large herbivores.

Indeed, scaling in accordance with elastic similarity also accounts for the apparent anomaly we observed earlier in the measured relationship between mammalian body mass and surface area. If we consider once again our animal composed of cylinders, we find that for any segment the surface area of the body is proportional to the product of length and diameter. From this it follows that the surface area is proportional to *(mass)$^{1/4}$ ×*

Horses have evolved as specialized running machines. Their ability to switch to different gaits, such as trotting or galloping, allows for the most efficient expenditure of energy at different speeds.

$(mass)^{3/8}$, which is $(mass)^{5/8}$ or $(mass)^{0.625}$. It will be noticed that this is the anomalous slope of the graph noted above.

Under the rules of elastic similarity, the implication is that the diameter of a body part should increase more rapidly and the length more slowly in larger animals than might otherwise be expected, and this has now been found to be the case in measurements as diverse as the bone size of dinosaurs and the chest proportions of apes.

On Being Large or Small

We can now begin to understand why bodily proportions and performance vary so greatly with size. The slender legs and springing gait of a small antelope become increasingly inappropriate for their larger cousins, whose bones, though shorter and stronger to bear the extra weight, still cannot withstand the stresses of leaping and bounding. This trend is strikingly illustrated by the massive bones and ponderous gait of elephants.

Although large size imposes limitations on athleticism, it also bestows many benefits in other areas. Perhaps the most important are increased efficiency and lower metabolism combined with reduced risk of predation and greater tolerance of environmental extremes.

An upper size limit is, of course, set by the mechanical strength of the body's building materials and physiological performance. Yet it seems that before these limitations are reached, ecological constraints and the reduced potential for rapid genetic change resulting from a greater interval between generations supervene to set an optimum size barrier for each group of animals. In periods of environmental instability and climatic change, selection will tend to favor smaller forms with short intervals between generations. In tranquil, unchanging times, large size will again become selectively advantageous.

At the opposite end of the size spectrum, animals face quite different problems. For locomotion, the two most important phenomena associated with diminishing size are the increasing significance of surface tension effects, and the shift in balance from inertial to viscous forces. This phenomenon is a manifestation of Reynolds number, which is described in Chapter 2, and reappears in discussions concerning the flight of insects and the swimming of protists.

INTRODUCTION TO ANIMAL DIVERSITY

The Earth was formed almost 5 billion years ago, and the earliest fossils yet discovered date back 3.5 billion years. From these simple beginnings life has proliferated and diversified into a bewildering array of organisms. Perhaps as many as 10 million species inhabit Earth today—and many more have become extinct. This range of size and complexity challenges our imagination, and to handle it a system of classification has been developed. This is a hierarchical arrangement in which organisms are ranked according to their possession of shared characteristics and common ancestry. Hence it is not just a method of pigeonholing species, but serves also as a mirror of their supposed evolutionary history.

At the highest level, organisms are divided into kingdoms. Today the traditional division of life forms into just two categories, either plants or animals, has been superseded by the recognition of five such kingdoms. These are the Monera (bacteria); Protoctista (algae, protozoans, slime molds, etc.); Fungi (lichens, molds and mushrooms); Plantae (mosses,

ferns, conifers and flowering plants); and Animalia (all vertebrate and invertebrate animals).

Within each kingdom species are united into genera, genera into orders, orders into classes and classes into phyla. The animal kingdom is grouped into more than 30 phyla, the members of each phylum united through common ancestry and a similar level of organization. These divisions represent the major brush strokes of evolution, and it is a humbling reminder of the enormous extent of the animal kingdom that the creatures with which we are most familiar—vertebrates such as mammals, birds and fish—comprise but a part of one of the smaller phyla, the chordates.

Almost 1.5 million distinct animal species have now been described (with perhaps more than six times as many remaining to be discovered), and fully 95 percent of these are invertebrates. The organisms that we shall meet on our journey are as strange and unfamiliar as the extraterrestrial bestiary of science fiction, but no less extraordinary. Lest the reader feel lost in the midst of this strange menagerie, be reminded that the principal characteristics of our players are explained in the Appendix.

Aquatic Origins

One characteristic shared by all phyla is an aquatic ancestry. Although some phyla today include representatives that have successfully adapted to life on land, all contain (or have contained) wholly aquatic representatives. It is for this reason that we shall begin our exploration of animal locomotion by examining the methods of propulsion that have originated in water.

In this context it is rewarding to reflect on the extent and diversity of Earth's marine habitats. Over two-thirds of Earth's surface—71.6 percent—is covered by sea. Animal life may be found at all depths, from the rich surface layers to the dark, abyssal depths of the Marianas trench 11 kilometers beneath the Pacific Ocean. As terrestrial creatures, we find it hard to comprehend the full extent of the oceans. Because more than 80 percent of them exceed 2 kilometers in depth, the volume of water available for exploitation by marine organisms is truly enormous. If the Earth's surface were to be smoothed out, the entire globe would be covered by water to a depth of 3,600 meters, a total volume of about 1.8 billion cubic kilometers. Even a single cubic kilometer is hard to visualize, but if spread out to a depth of only 1 meter, it would cover an area 32 kilometers square.

Given the extent of marine habitats and the length of time that they have been inhabited, it might be supposed that the bulk of the earth's animal species would be found here. Were it not for the extraordinary success of the insects, a group virtually without marine representatives, this would indeed be so. However, the insects outnumber all other forms of animal life combined by at least four to one, giving pride of place for diversity to terrestrial forms. This is a reflection of the enormous richness of terrestrial habitats.

The Elements of Evolution

It is sometimes thought that evolution proceeds ever upward, toward a clearly defined goal of increasing sophistication and intellectual superiority, crowned by mankind as its ultimate achievement. Linked to this notion is the belief that ranked beneath mankind in a descending hierarchy are lesser forms whose status is defined by their position in the table of classification. Nothing could be further from reality. The direction of evolution is random, and classification is an attempt to group together organisms that share common attributes in a way that, it is hoped, reflects the history of their descent from ancestral forms. To imply that one organism is inferior to another by calling it more primitive displays a deep lack of understanding of the nature of evolution. Each species represents a successful excursion into new ways of making a living. To be is to have succeeded.

The principles of evolution by natural selection put forward by naturalist Charles Darwin in his *Origin of Species* in 1859 are now universally accepted by biologists. All that is questioned are details about the relative importance of contributing mechanisms of selection and the rates at which they operate. Although evolution generally appears to be a gradual process, it is clear from the fossil record that from time to time there have been sudden and dramatic changes.

The genetic basis of inheritance ensures that the offspring resulting from any union will show variation. In certain situations, one variant has a greater chance of passing on its genes to future generations than does its siblings. Conversely, other variants have reduced chances of survival and die out without leaving descendants. Such unsuccessful gene combinations are in this way eliminated from any future role in evolution. Whether the variation is related to limb length, color pattern, resistance to toxins, tolerance of climatic extremes or a vast number of other factors, these slight changes will be selected and enhanced until isolated populations become so genetically distinctive that they no longer breed with the parent population. At this point the two populations have become separate species. The species living at any moment in time are the fortuitous survivors of innumerable selective processes, the ebb and flow of shifting genetic fortunes. At any point, selection must act upon the raw material to hand, seldom if ever being able to utilize ideal structures and substances. Evolution is the ultimate expression of successful compromise.

The Dawn of Competition

The early fossil record is tantalizingly sparse, yet there is evidence that by 1.4 billion years after the formation of the Earth, simple living organisms resembling bacteria had made their appearance. For the next 2 billion years and more, Earth's developing oceans and their growing populations of photosynthetic algae presented a scene of pastoral tranquility. Although there were few species in these communities, they occurred in great abundance, flooding all available habitats. There was little pressure for change, because competition was slight, and evolution proceeded but slowly. Then about 900 million years ago a new force appeared, shattering the calm of the oceans and dramatically accelerating the pace of evolutionary change. It seems probable that the great burst of diversification that resulted in the appearance of many new phyla in late Precambrian times stemmed directly from the advent of herbivorous protozoans. From such protozoans, feeding initially on bacteria and later on algae, it was but a short step from herbivory to carnivory, and life was never the same again.

Why should this shift in feeding habits have such far-reaching effects? The short answer is competition. The proliferation of new trophic levels and their associated food chains initiated an ever-escalating arms race in the hunt for improved methods of prey capture, defense, concealment and reproduction—precisely the environment in which natural selection flourishes. And flourish it did. Within a very short spell, perhaps only a few tens of millions of years, both multicellular plants and animals were firmly established in the oceans, which now seethed with life.

Many factors contributed to the growing complexity and diversification of living systems in Precambrian times, and it is pointless to attempt to rank them in order of importance. Certainly one of the primary influences on animal organization has been the constraints imposed by locomotory mechanisms, and it is this story that we shall be following in later chapters.

2

THE
MECHANICS
OF ANIMAL
LOCOMOTION

INTRODUCTION

In the previous chapter we examined some underlying biological principles, and it is now necessary to consider some basic engineering concepts. In science, words often have more precise meanings than they do in everyday usage. Thus weight and mass, for example, are frequently confused. Mass is an intrinsic, unchanging property of a body, measuring its inertia, or resistance to motion. It does not alter in different parts of the universe and it is not definable in terms of other physical variables. Weight, on the other hand, is an extrinsic property measuring the pull of gravity of a body and varies from place to place. Far out in space people and objects are weightless, while on the moon their weight is one sixth of what it is on Earth. Thus, weight is a force, meaning that it has both magnitude and direction, and can be defined in terms of the fundamental properties mass, length and time.

Similarly, a distinction exists between speed and velocity. Whereas speed is simply a measure of distance traveled in unit time, velocity possesses not only magnitude but also direction. The distinction is well illustrated by considering a man running for 100 meters along a straight track and then running back. If he does this in 40 seconds his average speed will be 5 meters per second. However, because his starting and finishing points are identical, his average velocity, or net change of position, is zero.

Another term we shall encounter is acceleration, which is the rate of change of velocity. A special case is acceleration due to gravity (g), the rate at which a body falls to earth. This varies slightly from place to place on the Earth's surface, but is approximately 9.81 meters per second per second.

The laws of motion propounded by Sir Isaac Newton (1642–1727) are still the foundation of modern mechanics. Newton's first law states that *a static body remains at rest and a body in motion continues at a constant velocity unless acted upon by an external force.* Thus, if there were no friction and no gravity, a moving object would continue to move in a straight line forever. To alter velocity or direction, therefore, it is necessary for an external force to be applied.

The second law states the relationship between force, acceleration and mass: *A force acting on a body gives it acceleration that is in the direction of the force and has a magnitude inversely proportional to the mass of the body.* Acceleration equals force divided by mass, so for a given expenditure of energy, large objects accelerate more slowly than small ones.

Newton's third law states that *whenever a body exerts a force on another body, the latter exerts a force of equal magnitude in the opposite direction.* Or stated another way, action and reaction are equal and opposite. Thus, people walk by pushing backward against the ground with their feet and in turn are moved forward by the reaction of the ground against which they push. Likewise, a fish moves by accelerating water backward and gaining forward momentum because of the water's reaction.

Marine animals that are sessile as adults typically have highly mobile planktonic larvae. This picture shows larvae of a sponge and a brittle star. Both swim by means of numerous cilia until they become too large, when they sink to the sea bed and metamorphose into their adult form.

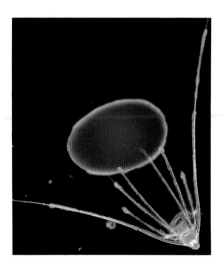

ENERGY

Energy is the capacity of a body to do work. For all practical purposes energy cannot be created or destroyed other than in atomic nuclear reactions. Thus no biological process involves the loss of energy but simply an exchange from one form of energy to another. Although energy cannot be destroyed, it can be wasted. This happens when it is converted into an unusable form. For example, muscles convert chemical energy acquired from food into mechanical energy, but this conversion is not very efficient and much of the energy is lost in the form of heat.

Mechanical energy may be divided into two categories, potential and kinetic. Kinetic energy is the energy of movement, while potential energy represents work already done but kept in storage. Walking up stairs involves the acquisition of potential energy. Falling from a window converts that potential energy into kinetic energy. From this it is clear that if work is done raising an object, for example a bird flapping its wings to gain altitude, it gains potential energy. When the bird stops flapping its wings and begins to glide earthward, it is converting the potential energy it has gained into kinetic energy.

Expressed scientifically, if a body of mass m is lifted through a height h, it is subjected to mgh work and so gains this amount of gravitational potential energy. If the same body now falls, its kinetic energy at velocity u is equal to $\frac{1}{2}mu^2$.

In order to climb a flight of stairs, thereby increasing one's potential energy, the leg muscles have to do positive work. But notice that in descending the stairs (assuming one walks rather than falls) the muscles have to do negative work.

MOVEMENT IN FLUIDS

The science of fluid mechanics developed during the 18th century through the advanced mathematical analysis of the laws of motion put forward by Sir Isaac Newton. Classical hydrodynamics, based on the concept of an "ideal" fluid, became a favorite arena in which mathematicians could display their virtuosity. However, classical hydrodynamics is largely an abstract exercise, and for many years it was of limited value to engineers, who had to deal with problems in the real world. Today the differences between real and ideal fluids are properly understood and it is now possible to integrate the largely empirical findings of engineers with their underlying theoretical basis.

For biological purposes, the mechanical properties of air and water may be regarded as very similar, and the principles discussed below are equally applicable to both media. It is only as the speed of sound is approached, far beyond the range of animal movement, that air's compressibility makes its behavior significantly different from that of water.

Leeches move about on the surface of their prey by looping and attaching themselves with powerful suckers. Piscicola *attack sticklebacks and other small fish. Many species are also able to swim strongly using up-and-down undulating movements.*

*Jacks (*Seriola*) are fast, powerful swimmers, and possess many features associated with this life-style. These include the form of the fins and the extreme narrowing of the body just in front of the tail. This shoal is feeding on fry.*

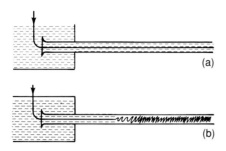

Fig. 2 The Discovery of the Reynolds Number. A thin stream of dye is introduced into water flowing through a tube. Below the critical value (a) the flow remains laminar. Above the critical value (b) turbulence sets in and the dye mixes with the surrounding water.

Hummingbirds are the most specialized flyers of the bird world. Their wing movements generate lift in both directions, and in many ways their flight is more like that of insects than of other birds. Their remarkable control allows them to perform elaborate courtship rituals on the wing, as seen here in the thorntail (Popelairia).

In a moving ideal fluid the flow about an obstruction is symmetrical and orderly. Such a flow is called laminar. In marked contrast, the flow within a real fluid soon ceases to be laminar and degenerates into turbulence. The principle underlying this transition from laminar to turbulent flow was discovered in 1883 by Osborne Reynolds, an English engineering professor, and today bears his name—the Reynolds number.

By introducing a fine trace of dye into liquid flowing through a glass tube (Fig. 2), Reynolds found that at a fixed tube diameter, turbulence set in when the fluid reached a specific velocity. However, if the velocity was kept constant, turbulence would appear if the diameter of the tube was increased above a certain value. Finally, he found that if the product of velocity and diameter were kept constant, the transition to turbulent flow occurred if a property called the kinematic viscosity fell beneath a critical value. Kinematic viscosity is dependent upon molecular composition, density and temperature.

Thus Reynolds number is a measure of the relationship between size, density, viscosity and velocity, expressed as:

$$Re = \frac{Size \times Velocity \times Density}{Viscosity}$$

Put another way, it quantifies the ratio of inertial to dynamic forces in a particular situation, and as such is an important concept in understanding the locomotion of small organisms. An object 1 millimeter across moving through water at 1 millimeter per second has a Reynolds number of 1. Thus hummingbirds function at Reynolds numbers of around 15,000, wasps at 4,000, fruit flies at 200 and the smallest flying insects at 1 or less.

Introduction to Flight

The Greek philosopher Aristotle (384–322 B.C.) attempted the first serious explanation of flight as part of his proof of the impossibility of a vacuum. Unfortunately, the Greeks' abilities in applied science did not match their considerable skills in mathematics. By relying wholly on theory and disdaining practical experiment, Aristotle got things badly wrong. In contrast to Newton, who some 20 centuries later clearly demonstrated that a body will continue in motion until a force is applied to stop it, Aristotle argued that a body, such as an arrow, could only remain aloft for as long as force was being applied to it. Once the force stopped, the arrow would fall to Earth. Thus, the Greeks believed that the air sustained flight by rushing in to fill the vacuum behind the arrow, so pushing it along. This interpretation was to remain unchallanged until Leonardo da Vinci (1452–1519) recognized that air is not an assisting medium, but a resistive one.

Even though we spend our lives surrounded by air, we are not really familiar with even its basic mechanical properties. Until we bicycle against the wind or witness the destructive force of a hurricane, we do not notice the pressure that can be imparted by moving air. Indeed, one of the great problems in 19th-century science was that, according to classical hydrodynamics, air should possess no resistance, and yet experience clearly demonstrates this not to be the case.

This discrepancy between the behavior of real fluids and their theoretical counterparts was, for many years, a major stumbling block in fluid dynamics. Known as d'Alembert's paradox, it was first expressed by French mathematician Jean d'Alembert (1717–1783) in 1768.

Classical theory's inability to account for d'Alembert's paradox was eventually traced to its failure to account for viscosity, the property of resisting a change in shape. This delay in reconciling the behavior of real and theoretical fluids is believed to have held back the understanding of basic aerodynamics by 50 years. Consequently, until the Wright brothers conclusively demonstrated that heavier-than-air flying machines were a practical reality on December 17, 1903, there were many distinguished scientists, including the secretary of the Smithsonian Institution in Washington, who argued vigorously that machines could never fly.

Gulls possess long, narrow wings that are characteristic of species that soar for great distances on rising air currents. Sailplanes also have wings of this type. The alula, or bastard wing, is clearly visible in this picture.

Fig. 3 Airflow Around a Wing. The streamlines join points of equal velocity and indicate the direction of air movement. Airflow is fastest where streamlines are close together and slowest where they are widely separated. Pressure is low where airflow is rapid.

PRESSURE DISTRIBUTION
OVER WING

Low pressure

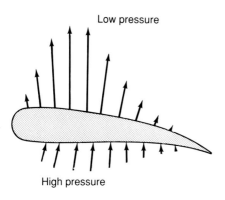

High pressure

Fig. 4 Pressure Distribution over a Wing. Aerofoils are designed to generate a substantial area of low pressure above and a smaller area of high pressure beneath. This reflects the spacing of the streamlines seen in this figure.

Bernoulli's Theorem

The velocity within a moving fluid flowing past a stationary object varies from point to point. If imaginary lines are drawn through the fluid joining points of equal velocity, a pattern of streamlines is generated. This is shown graphically in Figure 3. Where the streamlines lie close together, the fluid is traveling most rapidly, and where they are widely separated the flow is slowest. The interpretation of this velocity field about an object in terms of pressure was the work of the Swiss physicist and biologist Daniel Bernoulli (1700–1782), one of the fathers of aerodynamic theory. Bernoulli recognized that the pressure on a body in a moving fluid has two components, one static and the other due to motion. The dynamic pressure due to motion (expressed by $\frac{1}{2}\rho u^2$, where ρ is density and u is velocity), is that felt on the nose of a body where the impinging flow is brought to a halt. By the law of conservation of energy:

$$\text{dynamic pressure} + \text{static pressure} = \text{constant}$$

Hence, the static pressure is least where the fluid is moving most rapidly. This is an essential concept in understanding how a wing generates lift. Because air traveling over the upper surface of a wing has further to go, it must travel faster than air passing beneath. Bernoulli's theorem means that a region of low pressure is developed above the wing (Fig. 4) and it is this that provides most of the lift, even though it may not be intuitively apparent.

Boundary Layer and Turbulence

It might be assumed that a submerged submarine cruising at speed is surrounded by molecules of water that are continually rushing past its metal skin. In fact, this is not the case. The submarine travels within a thin envelope of static water, which isolates it from the moving water beyond. Similarly, a soaring eagle travels encapsulated within a bubble of stationary air. This static region close to a moving body is called the boundary layer. Although body and fluid are moving together at the same velocity where they are in contact, a short distance away the molecules are passing rapidly past one another. Thus, within the boundary layer there is a sharp velocity gradient with powerful shearing forces, which can lead to the formation of turbulence. Control of boundary layer turbulence is essential for efficient movement, particularly at high speed, and has exerted considerable influence on the course of evolution in many animal groups.

In essence, turbulence consists of the mob violence of innumerable, tiny, uncoordinated vortices. To create and launch such little whirlpools requires the expenditure of energy, which must be drawn from the motion of the body. Form drag is caused by turbulence in the wake and hence, like skin friction, is ultimately a consequence of viscosity in the boundary layer.

Turbulence, as we have just seen, appears in the wake, where its effect is far from beneficial in terms of efficiency. However, the situation is rather different within the boundary layer. If we examine flow in a boundary layer over a surface of some length, we find that at first it is entirely laminar, but at a certain point downstream it suddenly thickens and starts to grow rapidly. This transition point is where turbulence first sets in, and its position is determined by when the Reynolds number exceeds a certain critical value. Thus turbulence can set in if either the velocity or the extent of the surface is large enough. The turbulence mixes up the slow and fast moving elements, making the velocity more uniform over most of the depth. There still remains, however, a very thin laminar sublayer in contact with the surface, in which the velocity gradient is extremely high. It is this sublayer that determines whether a surface is aerodynamically rough or smooth. So long as surface irregularities (such as rivet heads on an aircraft) remain within the thickness of the sublayer, the surface is effectively smooth. This principle assumes considerable importance in understanding the wings of small insects at low Reynolds numbers, because the boundary layer increases in thickness as the Reynolds number becomes smaller.

Although the effect of roughening the surface will be to increase skin friction, in the case of a bluff body (i.e., one that is not streamlined) the total resistance may be reduced. In such a body, form drag is high because the flow separates to form a broad wake. If the point at which separation occurs is delayed, the size of the wake is reduced and the drag is less. In a turbulent boundary layer, energy is being constantly introduced in the form of fast-moving air fed in from the outside, replacing lost momentum. The overall effect is that the turbulent boundary layer gets further back along the surface before separating, so reducing the size of the wake and hence the drag.

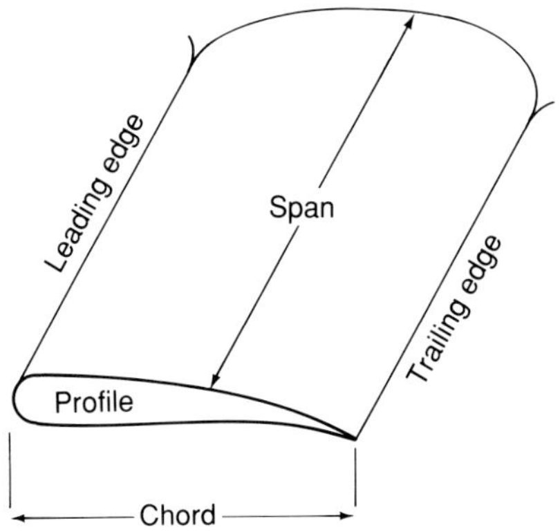

Fig. 6   The Parts of a Wing

The Aerofoil

If a flat plate is gently inclined in an airstream, forces acting on it are not symmetrical, but may be resolved into vertical and horizontal components of lift and drag (Fig. 5). Quite empirically, it will be found that the ratio of lift to drag is improved by bending the wing slightly to give it camber, or curvature relative to the chord line (see Fig. 6 for the terminology of wing parts). Further improvement is gained, also empirically, if the plate is rounded in front and tapered behind. Such a shape smooths the airflow and provides the best lift-to-drag ratio, and is typical of aerofoil surfaces such as aircraft wings.

To predict the airflow past a given aerofoil section is extremely difficult, but in 1905 the Russian mathematician N.E. Joukowski found an elegant alternative approach to the problem. He began with a rod of simple cylindrical cross-section, about which the airflow was essentially symmetrical and well understood. By using a mathematical approach called conformal transformation (similar to that employed by cartographers to make a two-dimensional representation of a spherical world) Joukowski was able to manipulate coordinates to generate shapes with predictable flow patterns.

In a Mercator projection of the Earth, angles remain unaltered but lengths become increasingly distorted toward the poles. This means that at the center of a Mercator map distances are accurate, but become increas-

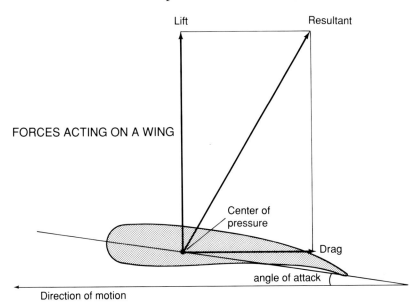

Fig. 5   Forces Acting on a Wing. The aerodynamic forces on a wing act through the center of pressure and may be resolved into vertical (lift) and horizontal (drag) components. The angle of attack, the inclination of the wing relative to airflow, influences profoundly the relationship of lift to drag.

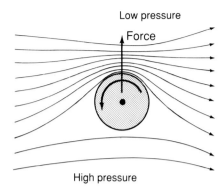

THE MAGNUS EFFECT

Low pressure

Force

High pressure

Fig. 7 The Magnus Effect. The airflow past a cylinder is symmetrical (apart from turbulence downwind). If the cylinder is rotating, as seen here, relative airflow is faster above (where the streamlines are closer together) than it is beneath. The resulting pressure differential subjects the cylinder to an upward force.

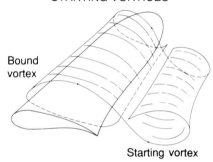

THE BOUND AND STARTING VORTICES

Bound vortex

Starting vortex

Fig. 8 Vortex Formation about a Wing. The wing is enveloped by the bound vortex. This is connected to the starting vortex, which is generated by air flowing off the trailing edge.

FORMATION OF WINGTIP VORTICES

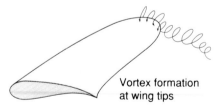

Vortex formation at wing tips

Fig. 9 Wingtip Vortices. High-pressure air beneath the wing is able to creep around the wingtips into the-low pressure area above. This movement of air creates a pair of vortices of opposite sense that, together with the bound and starting vortices, unite to form a single toroidal system.

ingly less so toward the edges. By using a mathematical approach employing the imaginary number $\sqrt{-1}$, Joukowski created a projection with opposite properties, namely one that became increasingly accurate toward the periphery. Hence he was able to confine distortion to the central region, corresponding to the innermost core of the wing section, where it had no significance. Later refinement of the technique now enables engineers to derive wing profiles precisely tailored to specific design requirements.

LIFT AND DRAG

When a body moves through a fluid the forces that act upon it, namely lift and drag, are related to the density and viscosity of the fluid, the size of the body and its velocity. This relationship was worked out by the English mathematician Lord Rayleigh (1842–1919) and can be expressed as:

$$\text{Force} = \tfrac{1}{2} \times \text{density} \times (\text{velocity})^2 \times \text{area} \times \text{constant}$$

The **constant** in this equation, which is known as either the coefficient of lift or the coefficient of drag depending on the forces involved, is a pure number whose value depends on the shape of the object and the Reynolds number of the motion. This fundamental relationship is of great importance in aerodynamics because once the coefficients of lift and drag are known it allows comparison of similarly shaped objects to be made regardless of size or velocity. Hence the performance of the wings of sparrows and eagles can be directly compared, even though the actual values for lift and drag may be vastly different.

Stalling

Stalling is a potentially catastrophic loss of lift that occurs suddenly and with little warning under certain flight conditions. Because it is a general phenomenon and not confined to specific types of wing, it may be analyzed through a study of the coefficients of lift and drag. When these are plotted against the angle of attack (α) (see Fig. 5), it is found that each behaves differently. Whereas the drag coefficient increases gradually as the angle of attack is sharpened, the lift coefficient climbs steeply to about 13 degrees. This is the critical incidence beyond which lift decreases rapidly because flow over the upper surface breaks away and becomes turbulent. In practice the equation defining the lift coefficient is usually turned around so that one speaks of stalling speed rather than stalling angle.

The Magnus Effect

In discussing d'Alembert's paradox it was pointed out that the flow of a moving ideal fluid past a long rod is symmetrical, indicating that the rod is not subjected to any net forces. However, if the rod is rotated, the fluid above and below it must move at different velocities (Fig. 7). From Bernoulli's theorem, differing velocities mean differing pressures, indicating that there is a net force generated in the direction of low pressure. Although this phenomenon was first recorded by Sir Isaac Newton in 1672, it is known today as the Magnus effect, after the German physicist Heinrich Magnus (1802–1870), who studied it in 1853. Golfers and tennis players will be indirectly familiar with the Magnus effect, for it is this, in conjunction with gyroscopic forces, that causes the strange flight of a sliced ball.

The Theory of the Wing

Box 1

The first person to recognize a unifying principle that united the Magnus effect with an aerofoil surface was the neglected British engineer F.W. Lanchester (1868–1946). So revolutionary was his theory that when it was first announced in 1894 nobody paid any attention, and 13 years passed before he was able to get it into print. Lanchester, recognizing the similarity of pressure distribution over an aerofoil to that surrounding a rotating rod, put forward the idea of air circulating about an aerofoil as its source of lift. This was a bold conceptual step, involving some rather esoteric consequences of classical hydrodynamic theory, in particular involving the principles of vortex formation.

First, Lanchester envisaged the whole wing enclosed within a horizontal vortex (Fig. 8). In such a system, the lift generated at right angles to the flow would be proportional to velocity, density and the strength of the circulation. Notice that, unlike the formation of drag, viscosity is not implicated in the creation of lift. Hence the disparity between theory and practice that gave rise to d'Alembert's paradox does not arise here. However, the concept of circulation leads immediately to two questions that stem from the theory of vortices. First, every vortex must be associated with equal and opposite flow, and second, vortices cannot simply begin and end. They must either attach to a surface or form a toroidal (doughnut-shaped) loop. How may these requirements be met? Lanchester's real genius lay in visualizing the circulation about a wing as part of a larger vortex system. With daring insight he proposed that the bound vortex circulating about the wing be treated (together with three other patterns of air movement described below) as part of a single large toroidal system. It is important to try to picture Lanchester's vision of the wing and its associated airflow because vortices now figure prominently in our understanding of animal flight.

Because the pressure on top of the wing is less than that beneath (as a consequence of Bernoulli's theorem), there is a tendency for air to try to creep around the edges. At the wing tips there is nothing to prevent this, which naturally gives rise to vortices that trail back from the tips of each wing, rotating in opposite directions (Fig. 9). The energy required to maintain these trailing vortices is manifested as induced drag, which together with profile drag makes up the total drag (see Box 1).

The bound vortex is thus linked to two trailing vortices from the wing tips. The picture is completed by an element rotating counter to the circu-

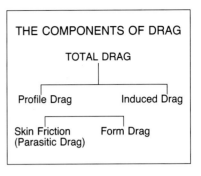

THE COMPONENTS OF DRAG

TOTAL DRAG

Profile Drag   Induced Drag

Skin Friction   Form Drag
(Parasitic Drag)

Modern jetliners have wings of considerable complexity, with a variety of flaps, slots and control surfaces. The flaps, shown extended in this picture, are used for landing and takeoff and generate increased lift when the aircraft is traveling at low speeds.

This shot of a red lory (Eos) coming in to land shows the ability of a bird's wing to open up so that each feather can act independently. This allows lift to be generated at slow speeds and at angles of attack far above that at which stalling would occur if the wing were solid.

Opposite page:
The short, broad wings of this African kite (Milvus) enable it to generate plenty of lift so that the bird can carry heavy prey. The widely separated feathers at the wing tips help limit the formation of vortices as high-pressure air beneath forces its way into the low-pressure area above.

The hovering flight of insects and hummingbirds has been likened to that of helicopters, but this analogy is oversimplified and misleading. The modern view regards animal flight as a succession of intermittent vortices rather than as a continuous system.

lation about the wing. Because it initiates and feeds the rest of the system, it is called the starting vortex (Fig. 8). As air first begins to move over an aerofoil the flow is irrotational, meaning there is no circulation. Two stagnation regions quickly develop, points at which the flow is brought to rest. One is where the air strikes the leading edge, and the other is just above the trailing edge. We know from Bernoulli's theorem that pressure is high in these two regions. As air from below endeavors to creep around the back of the wing toward the rear stagnation point, it is impeded by the pressure gradient and is curled back to create a vortex running the entire length of the trailing edge.

As it is washed off the wing, the starting vortex initiates a reaction in the opposite direction, and this is the circulation of the bound vortex. Thus the starting vortex, bound vortex and pair of trailing vortices effectively function as a total toroidal system. At every change of airflow, a new starting vortex is thrown off, generating a change in circulation and the lift forces.

This view of lift being derived from circulation comes from the study of rigid aircraft wings. While it seems probable that these forces are the ones that predominate in the steady gliding flight of animals, it is clear that the story is far more complex in flapping flight. Actual measurements show that lift several times greater than can be accounted for theoretically by steady flow conditions is generated by flapping flight. This is particularly evident at takeoff and in hovering.

Hovering flight of birds has, for some time, been explained by analogy with the rotor action of helicopters. Known as the momentum jet theory, this envisages the area swept by the wings as an actuator disc through which a mass of air is accelerated downward in a constant stream. The momentum imparted to the air provides an upward reaction to the bird. Such a model is applicable to the helicopter, but is inadequate when considering the reciprocating action of an animal wing.

A more acceptable explanation of flapping flight, applicable to both birds and insects, was put forward in 1979 by J.M.V. Rayner of Bristol University. This vortex ring theory envisages each wing beat as creating a small-cored toroidal vortex ring. These are shed to produce a wake of vortices, whose momentum sustains the animal. In hovering, the rings are circular and stacked coaxially, but in forward flight, the rings become elliptical and tilted. Rayner's explanation of the basic aerodynamic principles underlying bird and insect flight are convincing and now appear to be generally accepted.

We have now charted our course and identified the major features of the landscape against which the main action of our story will be set. We have looked at animal structure and function, considering the implications of size and the basis of evolutionary change. We have examined some basic properties of air and water and considered how mechanical forces exert their effects. It is now time to move in closer and start adding detail and texture to the scene.

3

LOCOMOTION IN WATER

INTRODUCTION

It is appropriate to begin our story in the sea, for it was here that life began. However, the earliest animals to move through the oceans were extremely tiny and subjected to mechanical forces that are beyond our experience and hence difficult to comprehend. So that we may begin our exploration of animal locomotion in familiar surroundings, we shall enter the evolution show late in the second reel and start by considering the way in which fish swim.

The muscles that power a fish act upon the axial skeleton and are arranged in blocks down either side of the body. As anyone who has eaten trout knows, the greater part of the body consists of white muscle, with a narrow band of dark muscle running along the middle of each side. Each type of muscle serves a different function. The dark muscle, which is permeated by capillaries and rich in mitochondria, gets its color from an abundance of myoglobin, an oxygen-carrying pigment similar to the hemoglobin in our own bodies.

In contrast, the white muscles derive their power by biochemical pathways that do not depend on the immediate availability of oxygen. Although liable to become fatigued quickly, over a limited period they are able to provide faster, more powerful contractions than the dark fibers.

The dark muscles, lying just beneath the skin in a position of prime mechanical advantage, are used for continuous, sustained swimming or cruising. It is an indication of the fish's economy of design that the bulk of its propulsion is provided by muscles occupying only between 5 percent and 20 percent of the body weight. The white muscle, in contrast, represents a luxury that no terrestrial animal, struggling against the burden of gravity, could possibly afford. A massive investment in energy and raw materials, amounting in some fish species to 80 percent or more of the entire body weight, this type of muscle is for sporadic use only, when short bursts of acceleration and high speed are required.

Intermediate muscle, a third category, combines properties of both white and dark muscle and is used by species that engage in extended bouts of rapid cruising. Thus fish may be regarded as possessing at least three forward gears, each powered by a different engine.

THE MECHANICS OF SWIMMING

Fish and all other swimming creatures propel themselves by invoking Newton's third law of motion and accelerating a mass of water backward. It is the various ways in which this is achieved that are of interest. Although a few species, such as the trunkfish (*Ostracion*), have rigid bodies and rely solely on fin movements for locomotion, most fish swim by undulating their whole body. The underlying mechanical principles, although applying equally to typical fish such as trout and goldfish, are most readily understood by considering elongated species such as eels.

When an eel swims, its body is thrown into a series of sinusoidal curves that sweep from head to tail, increasing in amplitude as they go. Each wave, as it progresses along the eel's body, displaces water in a characteristic manner, and it is the forces resulting from this displacement that impart forward motion to the eel. The same principle of water displacement applies to the Maltese boatman propelling himself by means of a single oar over the stern of his dingy and to most other fish as well, even though the undulations are less conspicuous in species with more conventionally shaped bodies.

Analysis of the continually changing waveform is complex, but we can understand what happens more easily if we consider just a short section of the body and resolve the forces on it at any one mome ̄ ̄to lateral and longitudinal components. This can be demonstrated by taking a typical fish such as a trout and introducing it to a board covered with evenly spaced wooden pegs. Despite the alien environment, the trout can travel fluently across the board. For the system to work the spacing between the pegs should be sufficient for three of them to be in contact with the fish's body at any given time. Figure 10 shows what happens. Pegs A and B impart sideways forces that cancel one another out and keep the fish pointed forward. Peg C is the one from which the fish derives its forward motion, and in mechanical terms equates to the backward displacement of water.

Bottom-dwelling fish like this turbot (Scopthalmus) *have undergone profound and asymmetrical modification. They lie on one side, making it necessary for one eye to migrate onto the upper surface. However, the change in orientation of the body has not greatly affected the normal undulatory swimming movements. The rippling motions of flatfish are anatomically still from side to side.*

VORTEX FORMATION

Not only does the experiment described above demonstrate the basic mechanical forces acting on a fish's body, but it may also reflect a deeper reality. It has been suggested that in swimming, fish generate a series of vortices along the body. These form just behind the gill region, increasing in size as they progress toward the tail. The positioning of the vortices is

Page 38:
The sinuous shape of this moray eel (Gymnothorax) *exemplifies the fundamental pattern of fish propulsion. The same mechanical principles apply to the movement of snakes and also microscopic flagella.*

analogous to the pegs on the board in Fig. 10.

Support for this relatively new and unexplored principle of fish locomotion comes from the study of a very different group of creatures. Among the smaller and less conspicuous inhabitants of the soil and its watery interstices are pale, thin nematode worms, such as *Rhabditis*. Like fish, *Rhabditis* propels itself by a series of undulations passing backward along the body. Small particles in the water close to the worm are seen to travel backward far more slowly than the waves of contraction pass along the body, which is not what one would expect. This is because they actually progress in a series of loops that demonstrate the creation of vortices forming at the crest of each undulation, and passing back along the body in synchrony.

Thus it appears that nematode worms and fish may indeed move by essentially similar means, the prime difference being that because of the small size of the worms and hence their effective Reynolds number (see Chapter 2), they leave no wake behind as do fish. It should also be emphasized that most free-living nematodes inhabit the rather special world of thin water films. These form a covering over soil particles, roots and such like, and provide an environment as yet little studied, and for which the undoubted special adaptations that exist are not understood.

ENERGY CONSERVATION AND CONTROL OF TURBULENCE

A major factor in evolution has been competition for ever-greater efficiency in the use of materials and the elimination of waste. An activity such as locomotion, which by its very nature involves a major expenditure of energy, has been a rich proving ground for such evolutionary experiments.

The drag that a fish experiences comes partly from the profile it presents to the water and partly from friction over its surface, which gives rise to eddies and turbulence. The shape of fish is very greatly influenced by streamlining to minimize profile drag, but relatively little is known of the ways in which fish control or exploit skin friction. In a number of fish species the skin is porous. Beneath it runs a network of fine canals filled

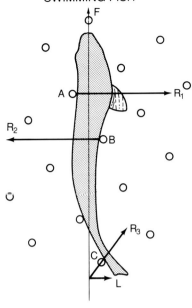

FORCES ACTING ON A SWIMMING FISH

Fig. 10   Forces Acting on a Fish. A fish can "swim" across a board covered with evenly spaced pegs. Pegs A and B provide lateral stability. The action of the tail against peg C generates a reaction (R3) that can be resolved into forward (F) and lateral (L) components. It is the element (F) that provides the fish with its propulsive force. (After Gray.)

In a typical fish such as a rainbow trout (Salmo), the sinusoidal movements of the body during swimming are less obvious than in fish with elongated bodies, such as eels.

with sea water. Although fish are not rapid swimmers in general, the Reynolds number at around which these species swim is close to the critical value at which smooth, laminar flow starts to disintegrate through the formation of turbulence. The porous skin with its underlying system of canals appears to iron out pressure irregularities by adding or absorbing water as necessary and so preventing the buildup of eddies.

A more widespread way of minimizing the fuel bill through the reduction of drag is by the use of slime. Many fish are covered by a layer of slime secreted from glands lying beneath their scales. Composed of long-chain molecules, slime controls turbulence in a way that is not fully understood, but the effect is dramatic. In barracuda, for example, the mucous film reduces drag by at least 60 percent, and is similarly effective in a number of other species of fish designed for rapid acceleration.

Even less well understood are the roles of the movable scales and projecting toothlike denticles found on many fast-moving fish such as Tuna. It has been postulated that they help to maintain laminar flow by introducing slight mixing within the boundary layer. This delays separation of the flow over the body and so impedes the buildup of turbulence. Spoilers on the wings of aircraft serve a similar function, but the mechanism is not yet properly understood in fish.

Although fish generally have shapes that are well streamlined and seemingly generate minimal drag, comparisons between scale models and living fish reveal that in life, fish are subjected to about seven times the amount of drag experienced by rigid models because of their undulating movements. This may explain why many fish move in a series of alternating cycles of active swimming and passive gliding. In the gliding phase, drag is reduced to the same level as that experienced by a rigid model, or even less, which makes this type of progression relatively efficient in terms of energy expenditure. A similar phenomenon is found in the flight of certain birds.

WAKE FORMATION

Traveling at the surface presents special problems. Admirals have long been aware that a submarine cruises faster and more economically when submerged. Beneath the waves the archconsumer of energy is drag resulting from friction over the surface of the hull. At the surface, however, these boundary layer effects are considerably outweighed by the energy consumed in the creation of a wake. The waves generated in the wake of an object traveling at the surface represent a considerable expenditure of work against the force of gravity.

The significance of this fact was not realized until after the construction of the liner *Great Eastern* in 1858. The largest ship ever constructed at the time, she proved to be disastrously underpowered, against all expectations of her designers. A subsequent investigation by a young naval architect, William Froude, unraveled the relationship between drag, speed and size, which is today known as the Froude number. Like its companion the Reynolds number, the Froude number is dimensionless, and plays a vital role in relating the performance of scale models to the full-size object, be it a ship or an animal.

For cetaceans, squid and fish traveling rapidly near the surface it actually becomes more economical to leap completely out of the water from time to time and proceed in a series of bounds than to cruise steadily and generate a wake.

THE DOLPHIN PARADOX

Fish are not alone in their need to control turbulence. Fast-swimming marine mammals too face similar problems, and for many years it has been thought that dolphins must possess special abilities in controlling boundary layer turbulence. Observations on the speed of swimming dolphins suggested that they were unduly energy efficient. In addition, it was noticed that dolphins generated far less turbulence than seals when swimming through waters rich in phosphorescent plankton. It was believed that transient ripples forming in the spongy fat layer underlying the dolphin's skin occluded local outbreaks of turbulence, damping it before it could spread. Despite extensive research by the U.S. Navy into the possibility of using compliant skins to improve the performance of ships and torpedoes, the early promise of energy conservation has not been borne out. Indeed, it now appears that assumptions about the speed and efficiency of dolphin swimming were probably erroneous. Although they are indeed capable of reaching speeds of 40 kilometers per hour, it is possible only for short bursts, and their normal cruising speed is about half this figure. It is highly likely that the original calculations were misleadingly based on dolphins riding the bow waves of ships, from which, like human surfers, they gain a substantial energy bonus.

The waves formed by the wake of a boat represent a substantial expenditure of energy against the force of gravity. Wake formation makes cruising at the surface an expensive form of travel for any organism.

EXPLOITING WAKE

As an animal swims, power is consumed in the formation of vortices, which are cast off to form a trailing wake. It should come as no surprise to learn that natural selection has found ways for this abandoned energy to be exploited. Indeed, several major features of fish morphology are now being recognized as directly linked to the manipulation of vortices generated in this way. In sharks, for example, the spacing of the dorsal fins, and probably of the tail fin too, is significant in this context. The leading fin creates a wake, which contributes to the overall drag, but this drag is less than if the fin formed a continuous structure along the whole length of the back. With each undulation of the body the trailing vortex is pushed slightly to one side. The spacing of the fins is such that at normal cruising speed each of these vortices is correctly positioned to be used by the following fin to gain impetus, a principle probably widespread among other groups of fish as well. An extension of the same principle explains the spacing of fish in shoals and the characteristic V-formation of flying geese and ducks. The wake of vortices formed by the leading individuals can be exploited by those following astern but only if they keep station accurately behind the leader.

Both birds and fish sometimes travel in a characteristic V-formation. By doing so, the individuals behind are able to exploit the turbulence generated by those in front, so conserving their own energy.

Dolphins (Stenella), *highly adapted swimmers despite their terrestrial ancestry, are often seen riding the bow waves of ships. This enables them to travel fast while saving a great deal of energy.*

FINS

For many fish, undulations of the whole body provide a good mechanism for sustained speed swimming. Most fish also possess a second method of locomotion for more delicate and subtle movements such as are needed for postural control, feeding and maneuvering through the complex structures of coral reefs. Such refined movements are provided by the fins, which may be divided into two classes, depending on whether they are attached along their length, or at a single articulation point. For some species, particularly those in which other factors predicate a rigid body, fins and tail are the sole propulsive structures. Thus porcupine fish are unable to undulate when inflated, and electric fish such as the mormyriform *Gymnarchus* have to remain straight and rigid for the proper functioning of their electric-field sensory system.

In the teleosts, or typical bony fish, the fin consists of a fan of skin supported by skeletal rays. The rays are articulated, which allows extensive movement, both along the body and from side to side. Indeed, when sudden acceleration is required and the tail gives a quick flip, the rays fold tight along the body, thus reducing drag. Muscles attached to the rays allow them, when erected, to be oscillated from side to side, sending waves along the length of the fin. By reversing the direction of these undulations, the fish is able to swim backward as well as forward and, by differential use of both paired and unpaired fins, maintain precise control of posture in all three planes of space.

For some fish, fins are the principal source of power for locomotion, rather than just a secondary one. The skates and rays, for example, have much-reduced body musculature and small tail, and have quite lost the ability to propel themselves by undulating body movements. Instead, the pectoral fins, which anatomically correspond to our arms, are powerfully developed and greatly enlarged to provide a means of propulsion that is both efficient and exquisitely graceful.

TAILS

In most fish the principal role of the tail is in acceleration rather than sustained cruising. The crucial factor in acceleration is the weight of water that can be displaced at a single stroke. A large tail will obviously move a greater mass of water than a small one, and it is therefore possible to deduce a great deal about a fish's mode of life by examining its tail.

It has been found that a 180-millimeter trout can acclerate from rest to a speed of 1.33 meters per second with just one double flick of its tail. Such speed is valuable for escaping predators or leaping up waterfalls, but is very demanding of energy and cannot be maintained for more than a few seconds. Even so, some fish are able to sustain substantial power output over prolonged periods. A 1.5-kilogram sockeye salmon has been measured ascending a river for about 1600 meters vertically in 5 days, covering a horizontal distance of 16 kilometers, without showing any signs of fatigue.

Tuna and some other fast-cruising fish have crescent-shaped tails attached to the body by a narrow stalk or peduncle. Powered by trunk muscles connected to tendons running through the peduncle, the tails of these specialized swimmers function as hydrofoils rather than by simply displacing water. Unlike those of other fish, the tails of tuna generate the main propulsive thrust in normal cruising swimming.

It has been demonstrated experimentally that the greatest drag is generated just in front of the tail, a region of the body that also produces virtually no thrust. For fish such as tuna that set a high premium on energy conservation, natural selection has gradually eliminated this expensive but nonproductive part of the body.

The shape of a leafy sea dragon (Phycodurus) *has evolved more for its effectiveness as camouflage than for its efficiency in swimming. Gentle fin movements are all that are needed to propel the fish slowly through the fronds of seaweed among which it lives.*

COMPROMISE IN DESIGN

Living as we do in an age of engineering excellence, we have become accustomed to a world of optimal designs; designs that utilize the best materials and perform their intended function with maximum efficiency and economy. But in biology there can be no perfection. Every organism, lagging behind the forces of selection that have molded it, exhibits compromise solutions derived from preexisting components. Moreover, most structures are required to perform a variety of separate roles, often with conflicting design requirements. For this reason most animals are generalists upon which certain specialist traits have been superimposed. This principle is well illustrated in fish, which have evolved in three different directions simultaneously. These are tendencies toward acceleration, maneuverability and cruising. The more adept a fish is in one of these skills, the less well it performs the other two. In general it is possible to deduce a fish's life-style by its morphology, linking performance to specific anatomical features such as body shape, fin structure or tail design.

Frequently, however, the factors affecting natural selection are not purely mechanical. Consider the pike (*Esox*), for example. Fierce predator of small fish, the pike relies on its tremendous acceleration for hunting success, and yet its body does not seem to meet the ideal design parameters for this type of swimming. For maximum thrust from a standing start, a deep body profile is predicted in order to minimize lateral movement, and yet the pike has a remarkably shallow profile in the front part of its body. Nevertheless, this seemingly lessened mechanical efficiency proves to be more than offset by the success in hunting that it bestows. It turns out that this is nothing to do with mechanics, but is a behavioral advantage. Small fish are far quicker to respond to the approach of a deep, predatory silhouette than to the approach of a shallow pike silhouette, significantly improving the pike's level of successful strikes compared to its competition.

Fish with rigid bodies, like this ornate cowfish (Arcacana), swim mainly by moving their fins. The tail is used primarily to gain acceleration during emergencies.

The pike (Esox) is a voracious predator of smaller fish. This one has caught a smaller member of the same species. Although pike may appear to our eyes to be shaped aggressively, to their prey their shallow profile renders them less threatening.

SNAKES

The undulating movement of a swimming snake is very similar to that of an eel, and the same mechanical principles apply. Most snakes and limbless lizards that swim do so only sporadically and show few special adaptations to life in water, apart from an increased stiffening of the forward part of the body. In fact, these animals face some peculiar disadvantages from their entry into an alien environment. Because they are air breathers and hence swim only partially submerged, they suffer increased drag through the formation of surface waves and they also have less of their body surface available for propulsion. Moreover, its sinuous body movements cause the snake's head to swing from side to side, making it difficult to approach prey in a straight line and strike accurately.

In contrast, the sea snakes show a number of interesting aquatic adaptations and, although air-breathing, can swim totally submerged. Not only can they move more economically by remaining wholly below the surface, so avoiding the creation of a wake, but their large flattened tails provide an increased propulsive surface. Special lungs and tightly sealing nostrils allow them to remain submerged for eight hours or more before surfacing, making them the veritable nuclear submarines of the serpent world. Sea snakes have also evolved a solution to the problem of how to approach their prey in a straight line, without swerving from side to side and so confusing their aim. The head and neck region is much reduced in size compared to the rear half of the body, giving greater mobility and significantly less inertia. As always, specialization incurs certain costs. Sea snakes pay for their aquatic superiority by being quite helpless on land, where the mechanical demands of locomotion, even for undulating animals, are quite different from those of life in water.

POLYCHAETE WORMS

Not all undulatory movements in water, however, are exactly the same. It might be thought, for example, that the same principles that govern snake and eel locomotion would apply equally to the serpentine movements of a swimming polychaete worm like *Nereis*. However, there exists an important difference. The long, segmented bodies of errant polychaetes bear a series of projecting paddlelike parapodia that have rather curious consequences for swimming. Unlike the waves of contraction that pass along the body of an eel from head to tail, those of *Nereis* run in the opposite direction, starting at the tail and progressing toward the head. How can waves traveling in this direction impart forward movement to the swimming worm but not to the eel? The answer lies in the cumulative effects of the parapodia. A smooth cylinder in water presents less drag if the flow is along its axis, rather than across it. The combined effect of the parapodia, however, is to reverse this situation and make the axial drag greater than the transverse drag. Under these conditions, the resultant forces that cause the eel to move forward become negative and so act in the opposite direction. For this reason the waves that propel *Nereis* forward have to move in the opposite direction along the body. We shall encounter this phenomenon again when examining the principles of flagellar movement. It should also be pointed out that in addition, each parapodium can be moved by a set of muscles that allows it to be used as an oar. These muscles will assume special significance when we come to consider the invasion of the land and the evolution of walking in arthropods.

OTHER UNDULATORS

A number of other invertebrates also swim using undulatory movements. These include large flatworms, whose bodies are thrown into graceful folds like the fins of a ray, as they dance through the water. The sea hare, *Aplysia,* has a pair of large, fleshy wings. These are lateral extensions of the foot, and as they flap up and down, undulations flow along the margins.

Squid and cuttlefish also make extensive use of undulating fins. Although they lack any rigid skeletal elements, their lateral fins provide a

The undulatory movements of snakes on land are also ideally suited to movement in water. Although lacking the refinements that make sea snakes so well adapted to life in water, European grass snakes (Natrix) like the one here are quite at home in ponds and streams.

flexible and highly controllable means of locomotion. They are able to cruise slowly, hover almost motionless, maneuver with precision or dart forward rapidly to catch prey, all with equal facility. They have also perfected a totally different method of locomotion, jet propulsion.

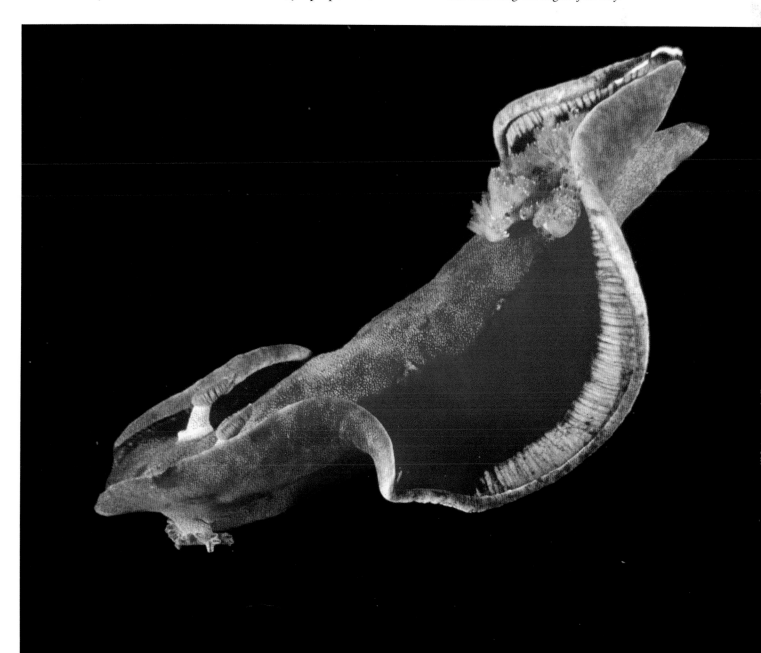

JET PROPULSION

The cephalopod mollusks have developed jet propulsion to a fine art. Octopus, nautilus, and above all squid and cuttlefish, have evolved jet systems whose level of sophistication far exceeds that of other animals. In order to understand how this method of propulsion works, we must remind ourselves of the key elements of molluscan morphology. Many familiar mollusks such as snails and oysters possess a protective armored shell. This is secreted by a portion of the body known as the mantle. Beneath the mantle in most groups of mollusks, including those which lack a shell, a secondary chamber, the mantle cavity, encloses and protects the delicate gills. Both the excretory and reproductive systems empty into the mantle cavity, as does the ink gland. Much of molluscan evolution and adaptation can be linked to specializations of the mantle cavity and its associated structures.

It is in the cephalopods that the mantle cavity reaches its most elegant form. Occupying the greater part of the underside of the body (Fig. 11), the mantle cavity opens to the outside by a groove that runs around the body where the edge of the mantle meets the head. Projecting downward from the mantle cavity there is also a muscular and highly maneuverable funnel, which anatomically equates to part of the foot in a snail. It is through this funnel that water is expelled for jet propulsion. The walls of the mantle cavity are well provided with muscle fibers, some of which are circular, some longitudinal and others radial in their orientation. Contraction of the radial muscles enlarges the mantle cavity, drawing water in through the large anterior groove. Once the cavity is full, the circular muscles contract, sealing the anterior groove like the flaps of a valve, and increasing the pressure of the water within. Normally the mantle expands and contracts rhythmically to draw a gentle stream of water over the gills, releasing it through the funnel. The muscular funnel can control the speed and direction of the expelled water with great precision. In the event of attack, the animal can dart backward at high speed, further confusing its assailant by adding ink to the expelled water.

Fig. 11   Squid Jet Propulsion. The squid draws water into its mantle cavity through a narrow slit behind the head and then expels it forcibly through a narrow, movable siphon.

SQUID JET PROPULSION

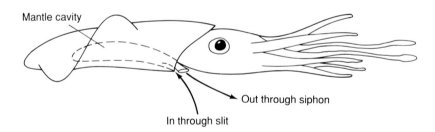

Mantle cavity

Out through siphon

In through slit

The cuttlefish (Sepia) is a cephalopod mollusk that is highly adapted for swimming. Precise buoyancy control allows it to float motionless. Gentle undulations of the membrane running along the body provide accurate positioning and posture, while an advanced jet propulsion system allows rapid movements for predation and escape.

The octopus is a close relative of the squid and cuttlefish. Suckers on its tentacles let it move fluidly over rocks. It can also travel with great rapidity by jet propulsion, as seen in this picture. When alarmed it can lay down a "smoke screen" by discharging special ink glands.

Squid

The jet propulsion system of cephalopods reaches its highest expression among the squid, some of whom can achieve speeds sufficient to project themselves well clear of the water and even to land 4 meters up on the decks of ships. Such a performance is made possible by a number of special adaptations that set the squid apart from other cephalopods.

As we saw in Chapter 1, most soft-bodied animals make use of a hydrostatic skeleton, with antagonistic sets of muscles acting on the fluid-filled coelom to generate rigidity through turgor, but in mollusks the coelom is much reduced. In squid this problem is resolved by incorporating a hydrostatic skeleton into the muscles of the mantle cavity itself, muscles which are further enhanced through the incorporation of collagen fibers. Reduction of bulk and the dead weight of a conventional skeletal system has greatly increased the squid's swimming performance, particularly its acceleration. Unlike the mantle cavity of the octopus, which has three sets of muscles acting at right angles to each other, and is infinitely variable in shape, squid have lost the longitudinal muscles, replacing them with a tunic of collagen. This not only helps to maintain body form, but also provides a firm insertion for the radial muscles as they contract to inflate the mantle cavity. The elasticity of the collagen also serves an additional function in maintaining a steady water pressure and hence improving control of the propulsive jet.

The circular muscles, which provide the pressure for locomotion, are divided into two types with different performance characteristics. Fast-twitch fibers are densely packed with contractile myofilaments, but have few mitochondria, which are the centers of cellular respiration. Although they fatigue rapidly, they can exert powerful tensions, which makes them valuable for brief bursts of speed. The slow-twitch fibers, on the other hand, possess fewer myofilaments but are rich in mitochondria and so function aerobically over long periods without fatigue. These are the muscles that allow the Japanese squid to migrate 2000 kilometers in two and a half months of continuous jet-propelled swimming.

Nevertheless, cephalopods are severely limited in their scope for significant improvement in performance. The very nature of their propulsive system imposes its own limitations. It can be shown mathematically that, for a given amount of effort, higher speeds may be achieved by slowly accelerating a large mass of water than by rapidly accelerating a smaller mass. Whereas fish can meet this requirement by fanning out and so enlarging their tails, cephalopods are limited by the volume of water that can be contained within the mantle cavity at any time.

Squid and cuttlefish are the descendants of a group of mollusks that dominated the Mesozoic seas 100 million years ago but were apparently unable to compete successfully against the growing onslaught of the bony fish. However, these modern cephalopods, despite their low profile, manage to hold their own remarkably well. Based on biomass, they certainly compete successfully with fish. In fact it has been estimated that the annual haul of giant squid consumed by sperm whales is in the region of 100 million tonnes. This significantly exceeds the annual weight of the entire world's commercial fisheries, which is about 70 million tonnes.

Bivalves

Cephalopods are not the only mollusks to make use of jet propulsion. Among bivalves, several genera of scallop *(Pecten)* and file shell *(Lima)* also employ this method of locomotion on occasion. As the upper and lower shells are drawn suddenly together, the mantle acts as a valve, trapping water within and allowing it to escape only through two small openings on either side of the hinge. As a result, the creature is driven forward, open-mouthed, as it were, in a series of rapid bursts, punctuated by brief pauses as the shells are opened wide to take in more propellant. This violent and exhausting method of propulsion cannot be sustained for long periods, and is only of use in avoiding the attentions of slow-moving predators such as starfish.

Bivalve mollusks like this file shell (Lima), *a relative of scallops, use jet propulsion to escape predators. Instead of forcing water out through the fringes of the mantle, as seen in this picture, a stream is directed on either side of the hinge.*

A curious feature of the hinge joint of bivalves is the absence of any muscles to open the valves of the shell. The shell is closed by contractions of the adductor — the muscle that forms the edible portion of a scallop — but there is no antagonistic partner, as in most muscle systems. Instead, the joint incorporates a compressible ligament made of a curious substance called abductin. Abductin is a naturally occurring rubberlike protein similar to the resilin that is found in the knees of grasshoppers and the wing insertions of flying insects. Abductin is able to store energy used in closing the shell, and release it with little loss of power to open the shell when the adductor muscles relax. This arrangement is of special importance to jet-propelled bivalves because their muscles are already working at much higher rates than those of most other mollusks, and hence need all the help they can get to generate the extra power necessary for swimming.

Scallops are preyed on by starfish, but can often escape by using jet propulsion. The power for this comes from muscles pulling the shells together. However, the shells open through elastic recoil of the hinge.

Other Jet-Setters

Jet propulsion has also evolved independently in a number of other groups of animals. For example, typical jellyfish swim by means of gentle pulsations of the umbrella, which keep them at their chosen depth, but with little directional control. More sophisticated are the siphonophores, colonial jellyfish relatives such as the Portuguese man-of-war *(Physalia),* in which individuals within the colony are specialized for particular tasks. Thus some members serve as gas-filled floats that control buoyancy, some for prey capture, some for digestion and others for reproduction. In *Stephalia,* for example, there are specialized medusae called nectophores that function

Sea turtles (Chelonia) swim with their front flippers, which act not as paddles, but as hydrofoils, using forces directly comparable to those generated by air movements over an aircraft wing.

solely as swimming bells to propel the colony through the open ocean by continually taking in and expelling water.

The other jet propulsionists are found among the insects. The immature stages of dragonflies, known as nymphs, pass their lives underwater as aggressive predators. They breathe through a series of rectal gills, which form a branchial basket occupying a barrellike chamber that fills the front two thirds of the rectum. Normally water is drawn in and pumped out gently to maintain a flow over the gills. But when necessary, water can be forcibly expelled from the rectum, propelling the nymph rapidly forward in a series of jerks. Curiously enough, this is their normal mode of progression, and is not just confined to escape from predators; but then dragonfly nymphs are not great travelers.

HYDROFOILS

Opposite:
Sea lions (Phocaretos) and their relatives are quite distinct from seals. They swim with their front flippers, using the hind flippers only for steering. Consequently the front part of a sea lion's body is powerfully built and they can move easily on land.

The heavily armored, rigid body of sea turtles sets these swimming reptiles clearly apart from the movements of their flexible snake relatives. They have developed a distinctive form of propulsion using their flippers. At first sight it might be thought that the turtle's flippers are used as oars for rowing, such movements being employed by a wide variety of animals. But upon careful examination the turtle's flippers are found to function quite differently, and are more analogous to wings in their action than oars. In cross-section a turtle's flipper resembles that of an aircraft wing, and in swimming is subjected to very similar forces. On the downstroke, the leading edge is tilted slightly forward, making a positive angle of attack to the water of about 25 degrees. The resultant forces include both vertical (lift) and horizontal (propulsive) components. On the upward recovery stroke

the angle of attack, and consequently the lift, becomes negative. In this way the vertical elements cancel each other out, leaving only a residual forward thrust.

The wings of penguins, which are not unlike the flippers of sea turtles, also function on the same principle. In fact, hydrofoil swimming is quite widespread among fast-moving marine animals. Consider the eared seals, which include the fur seals and sea lions. Unlike the true seals, which are awkward on land and which swim like whales and dolphins with vertical movements of their tail flukes, the eared seals are agile on land and use only their front flippers for hydrofoil swimming. This distinction is strongly reflected in differences between the skeletons of the two groups. In true seals, the bones of the front flippers and the associated pectoral girdle are not strongly developed, but the lumbar vertebrae, which are intimately involved in moving the hind flippers in swimming, are much enlarged. The contrast with eared seals is striking. Because these seals use their front flippers for both walking and swimming, the limb bones and shoulder blades are massive to support the extra musculature. Likewise the bones of the neck are also powerfully developed. California sea lions, which have been well studied, were found to generate thrust throughout the flipper cycle. Toward the end of the hydrofoil stroke, the flipper is turned broadside and briefly used also as a paddle. This combination of swimming techniques produces more power than either paddling or hydrofoil swimming could do on its own. Control of speed is primarily achieved by changes in stroke amplitude and angle of attack. Directional control, particularly in tight maneuvering, is aided by the hind flippers, whose role in water appears to be little more than that of a rudder.

Although relatively few animals possess flippers like those of turtles, the underlying principle of hydrofoil locomotion is quite widespread among fast-swimming animals. The flukes of whales and dolphins function in this way and so does the thick, muscular tail of the tuna, which has lost the ability to undulate. Indeed, in many fishes the distinction between undulatory and hydrofoil locomotion is rather blurred.

The harbor seal (Phoca) *swims by powerful up-and-down movements of the hind flippers. The shoulder girdle and front flippers are not well developed in the true seals, which despite their underwater agility are clumsy and awkward on land.*

ROWING AND PADDLING

The wings of penguins (Aptenodytes) are shaped similarly to the flippers of turtles and sea lions and they too function as hydrofoils.

Rowing and paddling are methods of locomotion that have evolved independently in many groups of animals, but the principal exponents are to be found among the arthropods. This is not surprising when it is realized that most paddling movements are little more than an extension of typical terrestrial walking movements. It might be argued that crustaceans, who make the most extensive use of paddling, have never gone through a terrestrial stage in their evolution. As bottom-dwellers, however, they were originally subjected to much the same mechanical forces as insects and other terrestrial arthropods. Indeed, it seems probable that paddlelike limbs arose as a direct consequence of the rigidity that arthropodization imposed on the body. This precluded the use of undulatory movements and necessitated the evolution of an alternative method of propulsion.

In the crustacea, virtually every appendage has been modified at some time or other to provide a propulsive surface by flattening and by the proliferation of fringed hairs. The structure of these fringed or plumose setae is quite elaborate, with the fringes themselves being further subdivided into smaller, very regular elements.

Primitive forms such as the fairy shrimp *Cheirocephalus* rely on rotary movements of the limbs on their thorax, but others such as the water fleas, ostracods and copepods use one or both pairs of antennae. The movements produced by beating the antennae tend to be very jerky, and generally intermittent rather than continuous.

Water fleas (Daphnia) *swim with their antennae, which are therefore large and spiny to resist the water. The legs are small and beat within the shell-like body to provide a respiratory current. Notice the small cluster of stalked protozoans* (Vorticella) *just above the eye. These feed through currents generated in the water by ciliary action.*

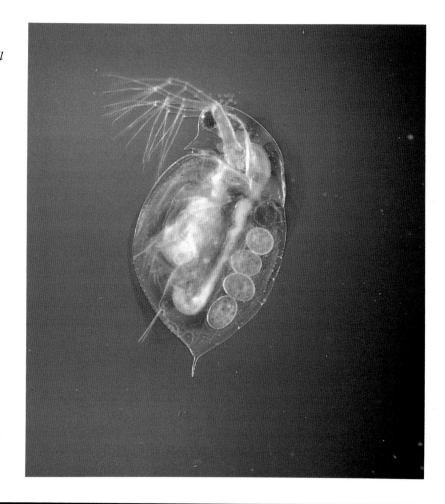

The limbs of this brine shrimp (Artemia) *are both numerous and complex. Like the limbs of most crustaceans, each is branched and clothed in long hairs to provide plenty of resistance to the water while paddling.*

The great size of planktonic copepod antennae, with their profusion of long setae, serves a dual function. Not only can the antennae move a substantial mass of water in relation to the size of the body, thereby providing good acceleration, but they also generate a great deal of drag when at rest, which slows substantially the rate of sinking. Other copepods, such as *Diaptomus,* rely entirely on their thoracic limbs for thrust. With the antennae folded out of the way, the five pairs of limbs beat in quick succession, each moving more rapidly than its predecessor. This allows *Diaptomus* to accelerate in one cycle from a standing start to a speed of 200 millimeters per second, which is remarkable for an animal less than 5 millimeters in length.

A special category of rowing is found among decapods such as shrimps and lobsters. This is in the use of the tail, with its extensible fan of paddles, for rapid backward escape movements. A very similar mechanism is found in the pupal tail paddle of midges and mosquitoes. Although short-lived, the acceleration of a large lobster is impressive, reaching as much as 9 meters per second. Sadly, for the lobster, it is this very ability to avoid danger that has become its undoing. The muscles powering the tail fan are the meat that make the lobster so tasty and have led to massive mechanized human predation, a threat that evolution has not yet managed to counter.

Among insects some of the most elegant rowing and paddling adaptations are exhibited by water beetles. What makes the water beetles' sophisticated aquatic modifications particularly interesting is that they are superimposed on an insect that is also well adapted for flight. The shapes of water beetle bodies have been strongly selected to generate minimum drag. Despite being heavily built, they manage to acquire neutral buoyancy by trapping an air bubble beneath

This freshwater shrimp (Gammarus) *illustrates well the basic crustacean body plan. Note the two pairs of antennae and serially repeated limbs on each segment.*

Prawns (Leander) *are also decapod crustaceans, and hence related to lobsters. Like them, they have a tail fan for rapid escape movements. In addition to walking over the sea bed, prawns can also swim by beating their paddlelike abdominal limbs.*

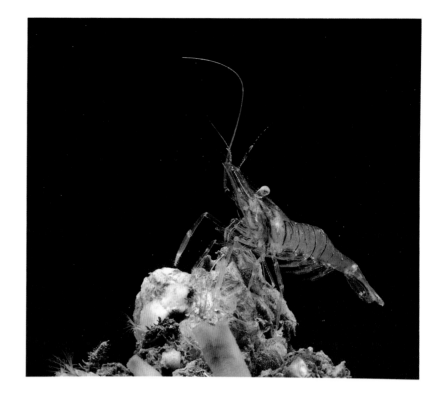

Water beetles (Dytiscus) *are powerful fliers as well as excellent swimmers. The air bubble protruding from beneath the wing covers helps this beetle maintain buoyancy while it paddles with its flattened, hairy legs.*

the wing covers, allowing all the power generated by the thrust of the legs to be allocated to forward propulsion.

Of the beetle's three pairs of legs, only the hind two are used for swimming, the front pair being stowed flush with the body in a shallow depression when not in use. Water beetle swimming legs are extensively modified to provide maximum thrust and minimal drag. This is achieved by substantially increasing the thrust-generating area for the power stroke and reducing it on the recovery stroke, which also allows the recovery stroke to be faster, so allowing a higher rate of paddling. The tarsal segments and the tibia of these legs are broadly flattened into paddles, and bear specialized swimming hairs. The trochanter and femur are also flattened, minimizing drag on the recovery stroke, but have no hairs. This is because, being close to the body, they move only a small distance and cannot contribute much to the power stroke.

The swimming hairs on the distal segments vary in form among different genera. Those of *Dytiscus* are long and thin, but in the whirligig beetles *(Gyrinus)* they are flattened, rather like the slats of a venetian blind. These hairs have special attachments that allow them to open out very easily to a predetermined point under the influence of water pressure, and then remain held there. When spread, the swimming hairs enormously increase the propulsive area presented to the water, but with low inertia and minimal use of materials.

The crayfish (Astacus), *a freshwater relative of lobsters, can swim rapidly for short distances by paddling with its expandable tail fan.*

The water boatman (Notonecta) *belongs to an order of insects quite distinct from the beetles. Yet like the water beetle it maintains buoyancy with the help of an air bubble, and its hind pair of legs is similarly modified for swimming.*

BUOYANCY CONTROL

Because all living tissues are denser than water, with bone, teeth and shells particularly so, aquatic creatures that are not bottom-dwellers face a potentially exhausting, long-term problem countering the forces of gravity. Ways of achieving buoyancy can be either dynamic or passive. Those creatures that rely on passive buoyancy compensate for their heavier components by incorporating an equivalent quantity of lighter material. Thus, some fish have a gas-filled swim bladder while others rely on the presence of low-density oils. For animals whose overall average density exceeds that of the water in which they live, dynamic countermeasures must be taken to generate the appropriate lift.

Small planktonic organisms such as crustaceans, jellyfish and many larval forms must swim more or less continuously to maintain themselves at the surface. But this becomes increasingly uneconomical for larger creatures. Squid, sharks and some bony fish make use of lateral fins to generate lift during forward swimming. For this to work, the upward forces must act in precisely the right place to allow correct posture to be maintained. In sharks, for example, the pectoral fins, which protrude just behind the gills, are solidly constructed, unlike the fins of most bony fish. Movement forward through the water generates lift on the pectoral fins in much the same way as lift is produced on the wings of an aircraft. Because the fins of the shark are placed well forward (Fig. 12), the lift they produce is applied some distance in front of the center of gravity. Without some compensatory mechanism, the shark would find itself progressing in loops. This is countered through lift generated by the upper lobe of the strongly upswept tail. In fact the lift produced by the fins varies in proportion to the velocity. In order to accommodate this and provide full control, sharks are able to alter the angle of attack of their fins. As in aircraft, if the shark's speed through the water drops below a certain critical figure, the fins suddenly lose lift through stalling. Hence sharks, and other fish such as tuna, which depend upon this type of lift to remain afloat, have to maintain a minimum speed through the water. Although this may appear wasteful of energy, the

minimum speed to remain afloat is also very close to the minimum needed to maintain a sufficient flow of water over the gills for respiration.

In squid, hydrodynamic lift is generated by the tail fins, which consequently force the front end down. Acting against this is an upward component generated by water being expelled from the mantle cavity through the siphon as part of the normal respiratory current.

Whirligig beetles (Gyrinus) *normally paddle on the surface too quickly to be seen as more than fast-moving ripples. High-speed photography shows them to be small relatives of the water beetle, with a similar body shape.*

The Art of Floating

History tells us that Archimedes, after a slight mishap with his bathwater, recognized that the buoyant forces acting upon an object immersed in a fluid were the same as the weight of the fluid displaced by the object. The density of fresh water is 1000 kg/m³. Seawater, because of its dissolved salts, has a density of 1026 kg/m³. Soft tissues such as muscle typically have densities of around 1060 kg/m³ and skeletal material roughly twice as much. Hence to achieve neutral buoyancy an animal must increase its volume with substances whose density is less than that of the surrounding water. As we saw earlier, most sharks are significantly denser than sea water, and must swim actively to avoid sinking. However there are some

FORCES ON A SWIMMING SHARK

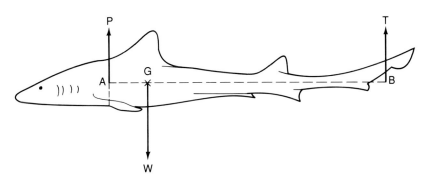

Fig. 12 Forces on a Swimming Shark. The weight (W) of a swimming shark in equilibrium acts vertically down through the center of gravity (G). Both the pectoral fins and tail generate lift (P and T) through points (A) and (B) respectively. Because (T) acts at a much greater distance from (G) than does (P), the tail needs to produce far less lift than the pectoral fins for the fish to maintain an even posture.

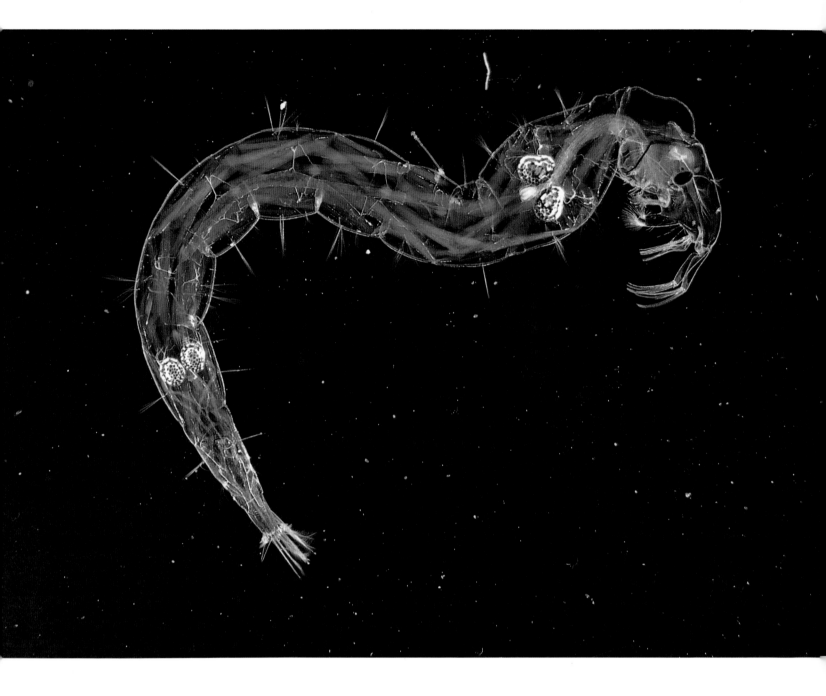

Many flies have larvae that are completely aquatic. The larval phantom midge (Chaoborus), a predator of water fleas, lies in ambush by hanging transparent and motionless in the water. It achieves neutral buoyancy through two pairs of gas-filled sacs, which are normally the only visible part of the body. Special lighting is required to make the larva as visible as in this picture.

species, the huge basking shark *Cetorhinus* among them, whose density is the same as seawater. These sharks all possess greatly enlarged livers containing a low-density oil called squalene, which occupies 30 percent or more of the total body volume. Waxy substances of similar density are found in a variety of bony fishes, including the coelacanth.

Another way of reducing density is to alter the composition of body fluids by substituting lighter ions (charged atomic particles) for heavier ones, while maintaining the correct osmotic balance. In general, this way of increasing buoyancy is not very efficient, and consequently it is mostly confined to smaller, soft-bodied creatures lacking shells or bony skeletons. The one notable exception is the chemistry of certain deep sea squid, which have replaced most of their sodium ions with much lighter ammonium ones. The density of the body fluids in such species is around 1010 kg/m³, compared with 1050 kg/m³ for the rest of the body tissues. With such a small difference in density, the volume of the fluids must be about one and a half times greater than that of the remainder of the body to achieve neutral buoyancy. It is, therefore, no coincidence that the body cavity of such squid is much enlarged and that the animals present a greatly distended appearance. Other creatures that successfully reduce the level of their sodium ions to maintain neutral buoyancy include numerous planktonic coelenterates, ctenophores and protozoans.

Swim Bladders

Evolution is progress by compromise; the exploitation of optimum solutions. For every roll of the evolutionary dice, the obvious benefits are usually closely matched by insidious disadvantages. Such a situation may be seen in the use of swim bladders and other gas-filled buoyancy systems. At first sight the benefits are substantial; maximum buoyancy for minimum volume being the most obvious. The drawbacks start to become apparent when we look at the swim bladders of bony fish. Because the density of gases such as nitrogen, oxygen and carbon dioxide is so much less than that of the body tissues, very slight changes in volume have a pronounced effect on buoyancy. Pressure changes resulting from changes in depth become highly significant and self-perpetuating. Descending only a fraction of a meter significantly increases the pressure on a fish. This will compress the swim bladder, making it less buoyant and tending to increase the rate of descent. To counter this the fish must expend energy, either by active swimming, or by secreting more gas into its swim bladder, a slow process. It is thought that swim bladders evolved from lungs, which in their original state are particularly unsuitable for buoyancy control. The walls of a lung are specifically designed to permit the rapid penetration of gases, and an increase of pressure within the lung will result in gas being forced into the bloodstream. The problem is compounded in fish by the presence of gills, which in turn pass the gases from the lung out to the surrounding water. To maintain buoyancy a fish would have to gulp air back into its lungs from the surface, clearly limiting its depth potential. In keeping with its origins as a lung, the swim bladders of certain more primitive fish still retain a connection to the mouth. Natural selection has wrought substantial changes in the course of swim bladder evolution. Surface area has been reduced, together with the originally extensive associated circulatory system. The walls themselves have been rendered almost impermeable by the incorporation of tight-packed guanine crystals. To control the composition and pressure of the swim bladder contents, a special glandular region has evolved on the swim bladder wall for the secretion or resorption of gas. Despite these improvements, swim bladders present their owners with limitations, mainly because the physiological processes involved in maintaining the requisite gas pressure are slow. A fish wishing to maintain neutral buoyancy must limit itself to a rate of descent of only about 2.5 meters per hour. Conversely, a fish adapted to a life in deep water will risk rupture if it attempts to swim to the surface.

There is no clear correlation between the depth at which an animal lives and its possession of a swim bladder, although such structures are less common among deep-sea species. The coelacanth *Latimeria*, which lives on the bottom at depths in excess of 400 meters, like a number of other deep-water forms, appears to have lost its swim bladder and replaced it with a fat body. Despite the limitations imposed by swim bladders, they are very widespread among fishes.

Sperm Whales

The marine mammals such as whales and sea lions face special problems. For them, buoyancy is provided by the lungs, which, as we have seen, are not well suited to this role. In fact most cetaceans remain submerged only relatively briefly and are actively swimming during this time. An exception is the sperm whale (*Physter*), which can descend rapidly to depths in excess of 2 kilometers and there hang motionless for half an hour or more until prey comes within striking range. The physiological problems posed by this life-style are considerable, and the sperm whale's adaptations are therefore particularly fascinating, but extremely difficult to study. Because they are great travelers, regularly migrating from the warm waters of the equator to the chill polar oceans, sperm whales' buoyancy control must function over a considerable temperature range. On a typical dive of 50 minutes in tropical waters, they can experience a temperature drop of 23 degrees C. This change in temperature, combined with the increase in pressure, has a substantial effect on the density of the surrounding water, and hence on the whale's buoyancy. The secret of the whale's ability to maintain neutral buoyancy over a wide range of depths and temperatures lies in the spermaceti organ and the oil it contains. Occupying about 80 percent of the head region, the spermaceti oil — up to four tons of it in a large male — provides considerable buoyancy. But spermaceti oil is unlike other oils in that it begins to solidify, and hence become denser and less buoyant, at temperatures below 31 degrees C. Unlike water, which freezes suddenly at 0 degrees C, spermaceti oil becomes increasingly crystalline over a range of several degrees. It transpires that the spermaceti organ is a massive heat exchange device to control the temperatures, and hence the density, of the oil. In addition to being richly supplied with blood vessels, the spermaceti organ is also traversed by one of the two nasal passages that link the blowhole to the mouth. While the left passage retains its respiratory function, the right passage has become highly modified. Its role is to pass controlled amounts of cold seawater through the spermaceti organ, countering the warming effect of the blood. In a typical dive, a sperm whale remains submerged for about 50 minutes, although dives of 80 minutes have been observed. This is followed by about 10 minutes of heavy breathing at the surface.

As the whale descends, its lungs collapse increasingly until at about 200 meters they cease to contribute significantly to the whale's buoyancy. In addition, cold water circulating through the spermaceti organ continuously cools the oil, so reducing its buoyancy. By the time the whale arrives at its hunting depth, it has achieved neutral buoyancy and can, in effect, turn off its engines and conserve energy.

In order to swim downward, overcoming the initial buoyancy of the lungs, substantial effort is needed. This raises the whale's body temperature significantly, which might be thought to pose something of a physiological problem. So it would, except that the whale has evolved an elegant way to exploit the excess heat. When the time comes for surfacing, blood courses through the spermaceti organ, transferring heat to the oil and liquefying it. The resulting increase in its buoyancy greatly facilitates the whale's return to the surface. Thus efficient use is made of the surplus heat generated in diving and an undue expenditure of energy is avoided.

And what benefit does the sperm whale derive from its elegant adaptations? The answer would appear to be related to the access it provides to enormous, untapped hunting grounds beyond the reach of competitors. The deep waters of the continental slope abound with squid, particularly the larger ammoniacal species, and these form the principal food of the sperm whale. It is interesting to note that the maximum depth at which a whale can achieve neutral buoyancy is directly related to its size. Small whales cannot swim so deep and are restricted to smaller squid. Only the largest whales — those in excess of 20 meters in length and a displacement of 50 tons — are able to penetrate to the depths inhabited by the giant squid.

It is ironic that one of the most elegant conquests of the ocean's depths should be achieved by an essentially alien life form. Like all other marine mammals, the sperm whales are descended from terrestrial ancestors and have only recently, in evolutionary terms, adapted to life in water.

Siphonophores

Gas-filled floats are not confined to fishes, but are also found among both mollusks and coelenterates. The siphonophores are a group of complex colonial jellyfish made up of specialized individuals, each performing specific tasks. Many siphonophores include pneumatophores, a single gas-filled "person" within the colony whose sole function is to maintain buoyancy. By controlling its gas content, the pneumatophore can often enable the colony to sink beneath the waves in inclement weather.

In the Portugese man-of-war, *Physalia*, the pneumatophore is enlarged into a massive float that stands high above the water surface. Far larger than necessary just to maintain buoyancy, it would seem, *Physalia's* pneumatophore consumes a significant amount of energy. Not only must it be kept filled with gas, but in calm weather it has to be flopped from side to side periodically to prevent its drying out. Its purpose appears to be locomotory, acting as a sail to propel the colony across the ocean surface at about

Although the hippopotamus can cover considerable distances on land in search of food, it normally lives submerged. Its eyes, ears and nostrils are aligned so they break the surface together. Being almost buoyant, the hippopotamus moves underwater with amazing ease and grace, expending little energy in the process.

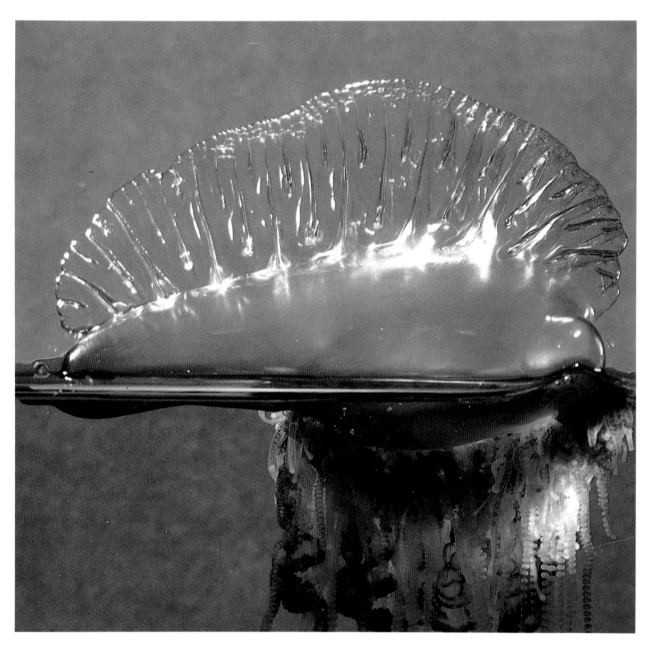

The Portuguese man-of-war (Physalia) is not a single animal but a colony of highly specialized, totally interdependent individuals. The gas-filled float is one such individual, providing buoyancy and acting as a sail to propel the colony.

45 degrees off the direction of the prevailing wind. When winds are light, *Physalia* pumps extra gas into its pneumatophore to create a larger sail area.

Nautilus

The most sophisticated use of gases for buoyancy is found in the cephalopod mollusks, a group of invertebrates that in many features, particularly their nervous system, is highly advanced. One aspect that sets the flotation mechanism of cephalopods apart is the role of the shell in providing a fixed volume system. This is in striking contrast to the swim bladder of fishes, which alters in volume with the slightest change in depth. Let us consider first *Nautilus,* which has a substantial external shell divided into a series of chambers. Writing in 1696, Robert Hooke (1635–1703), the English scientist, noted, "The animal has the power to fill or empty each of [the chambers] with water, as shall suffice to poise and trim the posture of his vessel, or shell, fitteth for that navigation or voyage he is to make; or if he be to rise, then he can empty these cavities of water, or fill them with air."

The living animal occupies only the last or outermost chamber of the spiral, adding successive new ones as it grows in size. Thus it is markedly different from the shell of a snail or other gastropod, which is totally lived in. The *Nautilus* shell grows as an equiangular spiral, each chamber successively increasing in size by 6.3 percent. Thus in a full 360 degrees there are 18 chambers, the outermost being three times the volume of the innermost.

The abandoned chambers in a *Nautilus* shell are sealed off by secreted calcareous septa. However, they remain joined to the living animal by a thin strand of tissue called the siphuncle. This runs through a porous calcareous sheath which perforates the center of each septum, and controls the gas content of the chambers. How this was done long remained a mystery, and the mechanism is only recently understood. As each new chamber forms, it is filled with liquid—the cameral fluid—which is gradually replaced over a period of weeks by a mixture of gases, predominantly nitrogen. Salts in the cameral fluid are withdrawn molecule by molecule through active transport across the cell membranes of the siphuncle. Thus the salt concentration in the cameral fluid gradually drops below that of the body tissues, creating a difference in osmotic pressure between the two. As a result, water is gradually drawn from the sealed chamber, to be replaced by dissolved gases, which slowly diffuse in. The process is not a rapid one, and it takes about a month for a new chamber to be emptied. Nevertheless, *Nautilus* possesses perfect buoyancy control, and is able to float without effort at any depth as it makes a nightly migration to the surface waters of the Pacific Ocean. This is because its buoyancy chambers are rigid and incompressible, and only slight changes need to be made to their gas content to control buoyancy.

This section through the shell of a pearly nautilus clearly shows how fresh flotation chambers are added as the animal grows. Notice how the shell thickness increases with size, and also the pore through the wall of each chamber. This is for the thread of living tissue that controls the gas composition within each chamber and adjusts buoyancy.

Despite its manifestly elegant buoyancy control, *Nautilus* is part of a failed evolutionary experiment. During the Jurassic and Triassic epochs some 200 million years ago, the seas were filled with a great variety of chambered cephalopods, including the hugely successful ammonites. However, by the end of the Cretaceous, some 65 million years ago, virtually all had become extinct — along with the dinosaurs. Locked for millions of years into one successful technology, they apparently lost their adaptive plasticity and were unable to meet the challenge of the emerging bony fishes with their swim bladders and improved mobility.

Other cephalopods, utilizing much the same basic principles, have managed to hold their own rather more successfully. In the cuttlefish *Sepia* the shell is reduced in size and covered by a fold in the mantle, so making it internal. Calcareous in composition, it consists of a porous matrix permeated by gas and fluid-filled spaces. There is no connection to the outside, and the gas — mainly nitrogen — diffuses into the interstices of the shell from the body fluids through water molecules being actively withdrawn by osmosis. The cuttlefish is able to regulate its buoyancy at any depth very precisely simply by controlling the proportions of gas and liquid in its shell.

LIFE AT THE SURFACE

The earth abounds in aquatic habitats, but there is one that demands our special attention. Separating sea from sky is the strange world of the water surface. A two-dimensional world of mysterious forces only molecules thick, it is nevertheless a habitat of enormous extent. In addition to the 22 billion hectares of water known to geographers, innumerable ponds, pools and puddles contribute to a vastly greater total.

Organisms that would exploit this in-between world face peculiar difficulties. For small creatures the principal problems relate to surface tension, which can trap and drown the unwary. However, with proper precautions the same surface tension can be turned to advantage.

The bubble raft snail (Janthina) *produces its own flotation system. This allows it to remain at the ocean surface, where it preys on jellyfish such as the jack-sail-by-the-wind seen in this picture.*

Within a fluid, a molecule is subjected to forces acting in all directions from neighboring molecules. For a molecule on the surface the neighbors, and hence the forces, are not uniformly distributed. This creates over the surface the effect of an elastic membrane, the surface-tension film. Numerous creatures have discovered how to overcome the problems of this in-between world and utilize the surface film for their own benefit. These include both aquatic and nonaquatic forms, depending on whether they move above or below the surface. For all of them, the critical relationship is the ratio of mass to surface area — in this case, the area in contact with the supporting surface-tension film. For larger animals particularly, this means finding ways of increasing the effective periphery of the contact area. This can be achieved by increasing the length and number of leg segments that touch the surface, or by developing appropriate out-pushings from the body.

Among the microscopic aquatic denizens of the essentially two-dimensional world are certain foraminiferans, marine protozoans related to the amoebas. They creep beneath the interface using a kind of protoplasmic streaming, capturing small planktonic organisms as they go. Larger inhabitants include a variety of predatory mollusks, which stalk the surface for jellyfish such as the Portuguese man-of-war (*Physalia*) and jack-sail-by-the-wind (*Velella*). The snail *Janthina,* too heavy to be supported by the surface tension unaided, secretes a raft of bubbles beneath which it hangs. The nudibranch mollusk *Glaucus,* however, does not require such artificial aids, despite its large size. Its body is drawn out into a series of fanlike lobes, which greatly increase the perimeter of its contract with the surface tension film, which is almost able to support the animal's entire weight. The difference is made up by a small bubble of air, which *Glaucus* swallows and holds in its stomach.

In freshwater ponds, hydras move across the surface film by looping, a method of progression more usually associated with inchworm caterpillars and leeches. They will also hang from the surface film with tentacles extended, fishing for small crustacean prey.

Many arthropods — insects and spiders — successfully live above the wa-

Glaucus is a strange and very beautiful nudibranch mollusk that lives in the surface film of the open ocean, where it preys on small jellyfish. Finlike extensions of the body help increase the effects of surface tension to keep it buoyant.

Mosquitoes (Culex) *have a special association with the surface film because of their complex life history. After passing the larval and pupal stages as aquatic, albeit air-breathing, creatures, they then emerge through the water surface, as shown here, to live as flying adults.*

ter surface, using the surface tension to prevent submersion. Most noteworthy are the water striders, whose long legs, clothed with water-repellent hairs, enable them to run with ease over the surface, scavenging for corpses and attacking any less well-equipped insect accidentally trapped there.

Far less conspicuous than pond skaters or water striders because of their small size, but actually far more abundant at the surface, are springtails. These tiny, primitive insects possess a jump mechanism that is really a terrestrial adaptation, but also works on water because they are so tiny. However, also because of their small size, springtails such as *Podura* encounter an unexpected problem. They cannot gain significant purchase on the surface, which makes them vulnerable to wind currents and also presents difficulties when they need to jump. To overcome this inconvenience, the springtails that live on water surfaces have evolved a special organ, the colophore, which serves as a wettable anchor and prevents them from slipping and drifting.

The upper size limit for this kind of life is reached by the large, long-legged spiders of the genus *Dolomedes*. Not only do they hunt like water striders for insects trapped in the surface film, but they also go fishing. Putting the tip of one leg through the surface and vibrating it in a characteristic way, they lure small fish such as sticklebacks within range and then, leaping into the air, dive through the surface to grasp them in their poison fangs.

In order to move over the surface in this way, insects and spiders have to increase substantially the area of their contact with the water. The extent of such contact may be readily appreciated on a sunny day by the distinctive shadows cast on the bottom of small ponds and streams as a consequence of the depressions created in the water surface by the velvety hairs at the tip of each leg. It is instructive to compare the posture and movement of *Dolomedes* with that of a large wolf spider, to which it is distantly related. *Dolomedes* does not walk on its tarsi, like terrestrial spiders, but flattens its legs to allow the long metatarsal segments to touch the surface as well. Then, instead of moving the legs on each side alternately as in walking, it swings both left and right legs simultaneously, as though rowing.

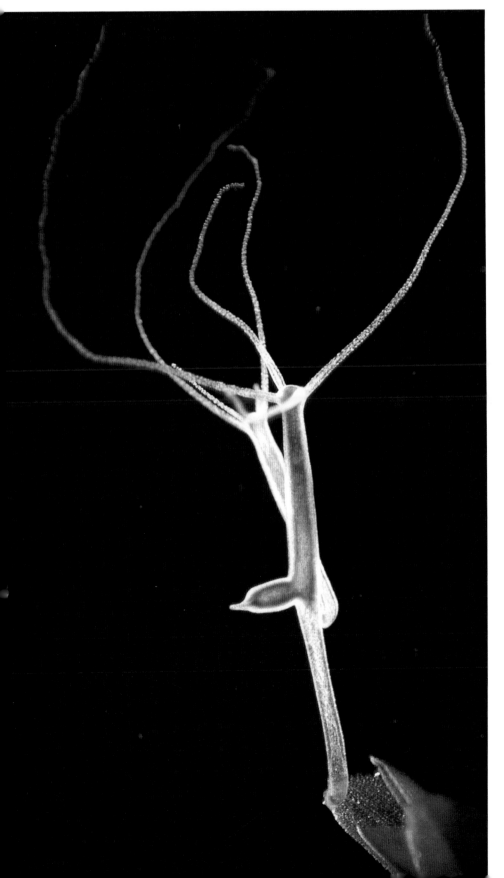

Hydras are freshwater relatives of jellyfish. They normally hang beneath the surface film and capture swimming water fleas and other small creatures with their tentacles, which are packed with cells containing poisonous darts. They are able to travel across the surface film by looping movements reminiscent of inchworm caterpillars.

Surfactants

Surfactants are chemicals that markedly reduce surface tension. They include a number of naturally occurring substances as well as man-made domestic detergents. Mammalian lungs, for example, contain dipalmitoyl lecithin to prevent surface tension from closing the alveoli, the tiny terminal air sacs where gaseous exchange occurs. In less sophisticated times children used to be amused by little boats powered by camphor. This acts as a surfactant to reduce surface tension behind the boat, which in consequence is pulled forward by the molecular forces in front.

Although the presence of numerous animals all pouring out surfactants would quickly render the water surface uninhabitable, it is curious that this method of locomotion does not appear to have been more widely exploited. Surface skimming by the controlled release of surfactants has been reported in only two groups of animals, both insects. One, the water crickets or ripple bugs (Veliidae), are close relatives of the pond skaters and

Perhaps the most abundant denizens of the surface film are springtails. These tiny insects (Podura) would be very susceptible to winds were it not for special adhesion organs in addition to the usual springtail jumping organ.

seem most appropriate candidates for this type of locomotion. But the others, a group of small rove beetles (Staphylinidae), are less obviously so. These beetles, belonging mainly to the genus *Stenus,* live in damp habitats close to ponds and streams, and not infrequently find themselves afloat by accident. Species living by still water use their legs to move across the surface film and regain the shore. In contrast, those living by fast-flowing torrents draw their legs to the side and apply eversible anal glands at the tip of the abdomen to the water surface, releasing a surfactant of unknown composition. Exercising a clear ability to orient and steer, they accurately skim back to shore and safety. Like many predatory insects, *Stenus* has well-developed eyes. Coupled with its known predilection for springtails, frequent inhabitants of water surfaces, it seems possible that this high-speed method of locomotion might well be used for the capture of prey.

Anatomically the *Entspannungsschwimmenorgan,* to use the delightfully descriptive German name for the eversible anal gland, appears to be derived from structures that in related staphylinid beetles are used for the production of repugnatorial secretions, part of nature's chemical warfare armory.

The raft spider (Dolomedes) *is one of the largest inhabitants of the water surface. It hunts both above and below, sometimes capturing small fish, which it drags onto nearby vegetation to eat. Tropical relatives also feed on frogs.*

Water striders (Gerris), and their close relatives the ripple bugs, are very common on the surface of ponds and streams, where they can move with great speed and feed on small insects.

THE MICROSCOPIC WORLD

Thus far we have been talking about relatively familiar animals — or at least ones that are easily seen. It is now time to pass beyond the limits of unaided human vision and examine how miniature life-forms conduct their lives. Very small organisms live in a curious world that is dominated by the effects of low Reynolds numbers, a world limited by viscous rather than inertial forces.

At Reynolds numbers less than 1, flow approaches that of an "ideal" fluid, with no separation and hence no form drag. For a microbe living in such an environment, the absence of inertia means that it cannot glide between strokes, but must swim continuously to keep moving. It is difficult to imagine what life would be like for us if we shrank into such a world, but it has been compared to swimming about extremely slowly in molasses — so slowly that no part of our body moved faster than the hands of a clock. Although this provides a good analogy to describe the forces to which we would be subjected, it would be quite wrong to assume that life for microscopic organisms is slow moving. Indeed, quite the opposite is true, as we know from Kleiber's law. With diminishing size, the pace of life increases.

The players in this section come from all parts of the animal kingdom, but the principal actors will be protists, unicellular (or more accurately acellular) organisms most of which are too small to be seen unaided. Because of their small size — some species are substantially less than one micron across — they are not easily studied and many details of their biology remain perplexing.

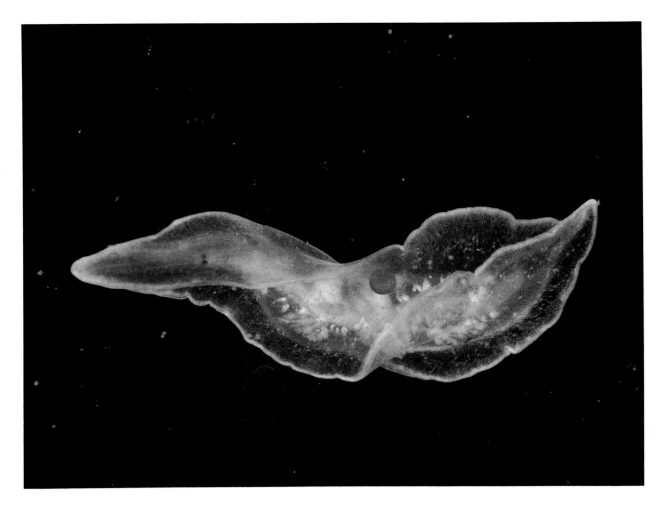

Flagella

Many protists and most motile sperm cells, including those of man, move by means of a long (10–500 μ) and slender (0.02–1.0 μ) undulating flagellum. Throughout the animal kingdom all flagella show essentially the same basic structure, which is remarkable and underlines the fundamental nature of the biological processes that occur within.

Ever since the uniform nature of flagella was recognized there has been speculation about the mechanism that enables them to undulate. Originally, it was thought the movement was passive, like the beat of a whip, with an impulse generated in the basal body passing along to the tip. However, the observation that amplitude increases rather than decreases as a wave progresses along the flagellum makes it clear that in some way power for the beat is being generated within the organelle itself.

We can now speculate with some confidence about what happens deep within the flagellum. Running its full length, and forming a central core, are two hollow fibrils, or axonemes. Arranged in a neat circle around the axonemes are nine other closely linked pairs of similar, hollow fibrils. One member of each of these outer pairs can be seen to have little arms reaching toward the neighboring pair. These fringed arms are molecular bridges similar to those formed between the sliding actin and myosin filaments of muscle, and are composed of the protein dynein. Motility of the flagellum results from the dynein arms attaching and reattaching to adjacent fibrils. This causes them to slide past one another like the filaments in a muscle fiber, and as in muscle, the energy for this comes from the transformation of adenosine triphosphate (ATP) into adenosine diphosphate (ADP). It is thought that the arrangement of nine outer fibrils and two inner ones, together with certain cross-linkages, provides the necessary mechanical sta-

Among the simplest multicellular animals are the unsegmented flatworms. Mesostoma moves slowly by means of innumerable cilia covering the undersurface.

bility to prevent the flagellum from buckling. This structure is remarkably consistent throughout the animal kingdom, and few significant variations are found. The principal modification to this basic "9 + 2" arrangement of fibrils is a doubling of the outer circlet to make "9 + 9 + 2." It is curious that this apparent increase in the capacity of the flagellar engine room is found only in the sperm of organisms practicing internal fertilization, whose needs for mobility are seemingly reduced. It is most unusual to find other departures from the "9 + 2" pattern of fibrils that characterize most flagella. However, among higher insects, such as flies, the sperm flagellum of some species is further modified by the loss of the axonemes and with differing numbers and arrangements of fibrils. Nevertheless, it is believed that the sliding mechanism of the fibrils remains essentially the same in all these variants.

Many nematode worms are microscopic. Ascaris is a large parasitic species that lives in the lungs of pigs, but nevertheless has the same body form and structure as those of its smaller relatives.

Mechanics of Flagellar Locomotion

We can understand how a flagellum moves by considering a conventional, elongate sperm cell with small head and long tail. Oscillations pass from the head, back along the tail, growing in amplitude as they go, driving the

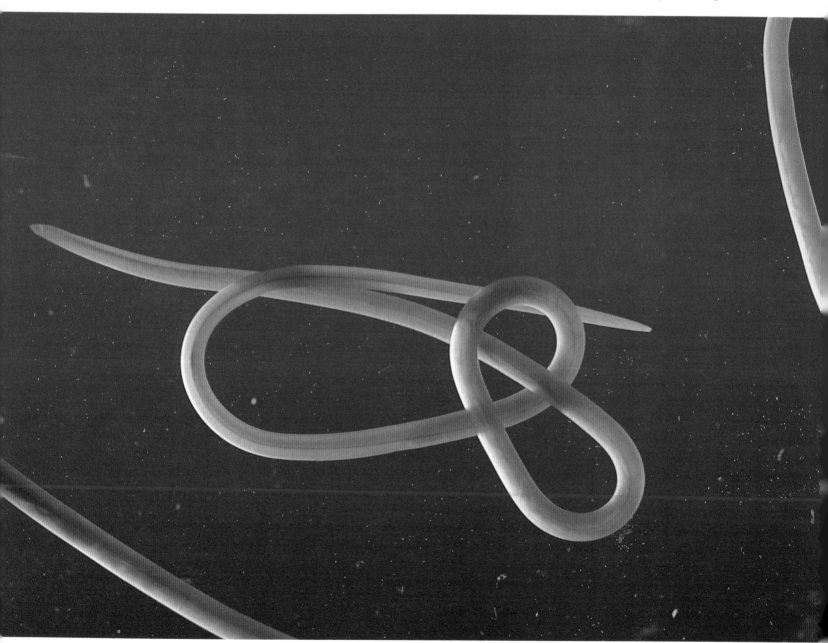

sperm forward headfirst. It can be shown mathematically that this forward motion occurs because drag along the flagellar tail (a product of length, velocity, sectional shape and viscosity at Reynolds numbers significantly less than 1) is only half that of the drag acting at right angles across the tail. When the forces acting on a beating tail are resolved, it is found that, as in fish, the lateral components cancel each other out, as one would surmise, leaving a positive forward component to propel the sperm. Relatively straightforward, it might be thought!

However, when we come to look at other organisms that rely on flagellar motion, we find some curious anomalies. In some species the wave passing along the flagellum travels in the same direction as the organism; in others it moves in the opposite direction. Some creatures are driven forward by a flagellum beating behind, others are pulled forward by a flagellum projecting out in front, and yet others are propelled by a pair of flagella projecting in front but wrapped around the creature's body to beat behind. The clue to understanding how these various arrangements can function is the presence on some flagella of mastigonemes covering the shaft. These extemely fine tassels are visible only under the electron microscope, and yet their influence is considerable and is analogous to that produced by parapodia along the body of a polychaete worm. Though small in themselves, their cumulative effect is to increase drag along the flagellar axis, making it about 20 times greater than the drag at right angles. This is the opposite condition to a smooth flagellum. Though it may not be immediately self-evident, because the axial drag is now greater than the transverse drag, the resultant force acting along the axis is negative. Consequently the flagellum and its owner are driven in the opposite direction.

Flagellar Power

Before leaving the flagellum, it is instructive to consider its power requirements and their implications. Why, for example, do we not find flagella or their derivatives driving larger organisms? The power needed to drive a straight rod (with a Reynolds number less than 1) is the product of its length and the viscosity of the surrounding medium times the square of the velocity. However, the undulations of a flagellum have been calculated to increase this figure fiftyfold. Although the size of the cell body being propelled clearly adds something to the flagellar fuel bill, because of the low Reynolds number, and hence the absence of drag, this is less than 20 percent of the total. Thus fully 80 percent of the power consumed goes to propel the flagellum itself, which is not what one might at first suppose. If we assume that the propulsive capability of a flagellum is proportional to its length, why are there no really long ones to be found? Quite simply because the drag is also related to length and increased length simply means increased power consumption without any gain in performance. Thus selection should, if anything, tend to act in favor of smaller, more energy-efficient flagella.

There is, however, one way in which the performance of a beating flagellum may be improved. Hitherto we have only considered a flagellum beating in a single plane, which is what happens with sperm tails. However, the tails of many protists are found not to move in just one plane, but to describe a helicoid or corkscrew action. Calculation reveals that provided certain mechanical requirements are met, this three-dimensional undulation can generate double the propulsive force of a flagellum beating in only two dimensions. For the system to work it is necessary that the front of the flagellum be secured and prevented from rotating, which is not the case in most sperm cells with small heads, but does effectively happen through inertia in the more bulky protists.

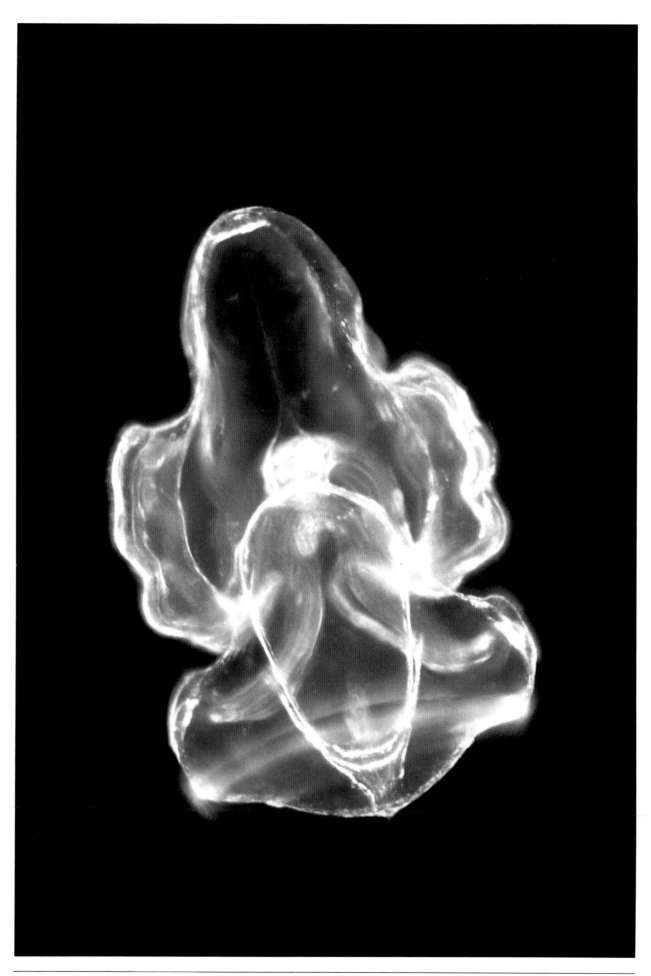

Bacteria

A notable exception to all that has been said hitherto about flagella is found in the bacteria. The flagellum of these organisms, which occupy a position on the evolutionary ladder quite distinct from either plants or animals, represents an earlier evolutionary experiment. Much smaller than true flagella, those of bacteria appear to be made of solid protein and lack the covering of a cell membrane. Many bacteria possess a number of flagella, which periodically aggregate together and function in unison during periods of activity, only to separate again when at rest. At first sight the movements of bacterial flagella are not markedly different from those of higher organisms, but this similarity is illusory. No wave is propagated along its length, not even a passive one. Instead the entire organelle appears to rotate in a series of clicks about a unique and quite remarkable linkage at its base that provides nature's closest approximation to a rotating wheel. A permanent kink in the flagellum creates the illusion of a propagated wave. Whereas the flagellum of higher organisms possesses a complex internal structure and the ability to generate within itself the necessary power for beating, these features are completely lacking in the bacterial flagellum.

Ciliary Movement

The second principal method of protozoan locomotion depends upon the cilium, which is sometimes thought of as a small flagellum because both possess a similar arrangement of internal fibrils. But this view of cilia obscures important functional differences. Although possessing the same "9 + 2" arrangement of axonemes and associated outer fibrils, the actual mechanics of ciliary movement are so distinct that they must be treated quite separately.

Cilia are distinguished from flagella by their much shorter length (5–12 μ), their whiplike movement and their occurrence in large numbers rather than singly or in small groups. Whereas a flagellum has a symmetrical beat, with several waves passing along its length at any moment, the beat of a cilium is markedly asymmetrical. Each beat is a separate incident occupying the entire length of the cilium, which in consequence has both a powered propulsive stroke and a feathered recovery stroke in each cycle. An important consequence of this different method of beating is the direction of water flow it produces and hence direction of thrust. Whereas the beat of a flagellum generates thrust, either negative or positive, along its axis, the thrust of a cilium acts at right angles to the axis and parallel to the body surface. Thus adjacent cilia, beating in unison at 10–20 hertz, enhance one another's efforts and result in the production of substantially greater thrust than can be generated by a few flagella. In fact those creatures that have adopted ciliary locomotion travel at one or two millimeters a second, which is approximately ten times better than flagellates can manage.

In practice cilia do not beat strictly in unison, but very slightly out of phase with one another. This metachronal rhythm, as it is called, is what makes ciliary movement so striking to the eye, when viewed through the microscope, presenting the appearance of wind blowing across a field of wheat.

The large protozoan phylum Ciliophora — usually referred to as the ciliates — consists of organisms bearing large numbers of cilia. In some species they cover the whole body surface, while in others they are confined to selected regions, often being associated particularly with the production of water currents to carry in food particles. The precise mechanism controlling coordination between adjacent cilia is not understood, but is closely associated with the kinetodesmata, a network of fibers running just beneath the body surface uniting each cilium with its neighbors.

Opposite:
This tornaria larva of a burrowing acorn worm swims actively in the plankton by means of bands of cilia, which as the animal grows become increasingly convoluted in order to support the extra weight.

Most protozoans are microscopic, but this giant foraminiferan (Globigerinoides) measures 30-40 mm across, making it one of the largest. It lives in the plankton, where it feeds on small crustaceans trapped in the protoplasmic covering of its spines.

Like flagella, the power output of cilia is limited, but because they are present in large numbers, their total effect can be appreciably greater. For this reason we find cilia propelling creatures rather larger than protists, albeit somewhat slowly. Metazoans propelled by ciliary action include rotifers, flatworms and ribbon worms, as well as the motile larval stages of all the aquatic phyla, particularly those with sessile adults.

We saw earlier that no benefit can be derived by enlarging a flagellum because the power needed to propel it increases proportionately with length. The same limitations apply to cilia (and also they would cease to be cilia if elongated to the length of a flagellum). However, cilia may successfully fuse together, producing stubby motile structures called cirri. These are found in some ciliate protozoa, where they can take a variety of forms. In some species, such as *Stentor,* they beat like enlarged cilia to generate feeding currents. In others the cirri themselves fuse together, producing an undulating membrane, which can be used for both feeding and locomotion. Some of the most specialized ciliates, such as *Stylonychia,* have large cirri confined to the underside of the body. These are used as little legs for a curious kind of tiptoe walking. In *Halteria,* another ciliate, the body is surrounded by eleven long bristles, which are apparently modified cilia. Sudden movement of these bristles causes *Halteria* to progress in a curious bouncing fashion.

Comb Plates

The most striking use of cirri for locomotion is encountered in the comb jellies, or ctenophores. These are distant relatives of the jellyfish, but sufficiently distinct to be placed in a separate phylum. The name *comb jellies* refers to their organs of locomotion, eight bands of combs encircling the body. The comb rows are made up of a series of plates, each bearing a neat row of large cirri composed of large numbers of cilia fused together. Beating metachronally, the comb plates of *Pleurobrachia,* the sea gooseberry, and other ctenophores flash beautiful interference colors.

In fact, comb plates are unusual because they do not beat regularly, but need to be triggered by currents set up by the beat of an adjacent plate. Although ctenophore comb plates appear to function like the serried ranks of oars projecting from a Greek trireme, their combined effect resembles jet propulsion rather than rowing. Water trapped by the combs is forced against the body wall and squirted out sideways. The ctenophores are the largest animals to make use of ciliary action in locomotion. However, they do not rely on it exclusively. In Venus's girdle *(Cestus),* a large species that may reach 1.5 meters in length, swimming occurs through gentle undulations resulting from the contractions of muscle cells. *Cestus* uses its comb plates solely for orientation.

Cilia are distributed far more widely through the animal kingdom than are flagella, which are confined to protists and sponges and, among the metazoa, to sperm cells and certain primitive excretory structures. In every phylum with marine representatives there are species with planktonic larvae. Living in the upper waters of the great oceans, these larvae drift at the mercy of the currents until the time is right to settle and metamorphose. However, they are not totally helpless. Bands of cilia enable them to make limited movements, which are used for vertical migration. Many larvae and other plankton pass the day at depths of several hundred meters, only swimming to the surface after dark. The purpose of this vertical migration cycle is not wholly clear, but it does allow a planktonic organism to play the currents and remain in more or less the same place relative to the seabed. As a larva grows and becomes increasingly heavy, so it finds it harder and harder to stay afloat. To compensate for this increased weight, the bands of cilia surrounding the body increase in length, becoming ever more convoluted, until there is no further room for expansion and life on the bottom becomes inescapable.

The role of cilia is not confined to locomotion, however, particularly in higher animals. Cilia are widely used to move mucous films with trapped debris, for example in annelid and molluscan feeding or in cleaning the respiratory surfaces of lungs. But the most surprising role of cilia has nothing to do with movement. Outgrowing their locomotory origins, cilia are also to be found throughout the animal kingdom performing new and quite unexpected sensory roles. Among the arthropods, in particular, with their tough outer covering performing the dual function of skeleton and protective armor, modified cilia serve as olfactory, acoustic and other sense organs to provide the nervous system with news of the outside world. Among vertebrates and higher invertebrates, light receptors within the eye are formed from ciliary structures that have undergone extensive evolutionary modification.

This completes our survey of the swimmers and drifters of the animal kingdom. From surface waters large and small we now descend to the ocean floor, where we shall start to prepare for the dramatic transition of life from water to land. For this momentous move, all the elegant adaptations and refinements wrung from the clay of life and honed to perfection over millions of years in lakes and oceans will count for naught. A new chapter with quite different problems and solutions is about to materialize.

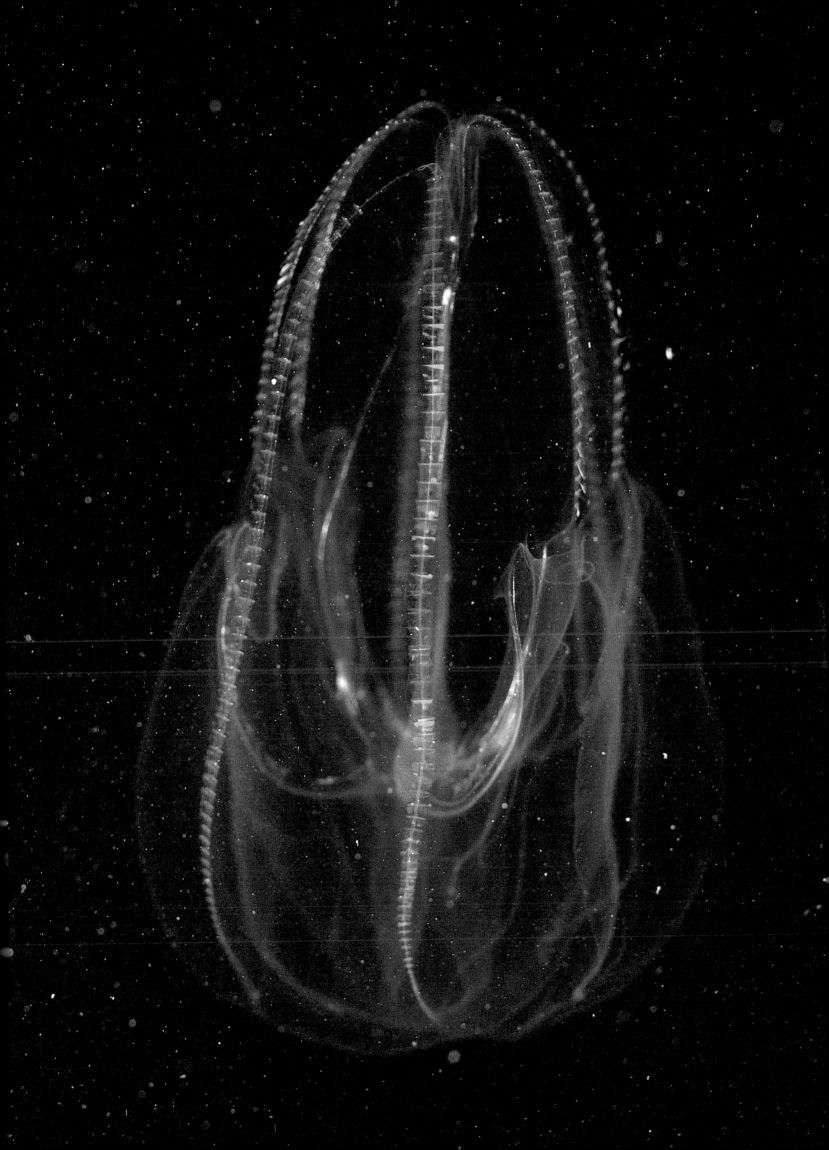

4

LOCOMOTION
ON LAND

Page 86:
There are many species of essentially quadrupedal animals that from time to time adopt a bipedal stance. The gerenuk (Litocranius) is a long-necked African antelope that spends much of its time standing on its hind legs as it feeds beyond the reach of other species on low trees and bushes.

It is now time to direct our attention to animals that move not by displacing water, but by pushing directly against a solid substrate. Although this will eventually lead us to familiar running and jumping terrestrial animals such as cheetahs, horses, fleas and frogs—and ultimately to humans—we shall begin this chapter by looking at the creatures that dwell on the sea floor. From here we shall be better able to understand the significance of what occurred when aquatic organisms first started to invade the land.

LIFE ON THE SEABED

Early inhabitants of the oceans, depending on sunlight for their energy, were confined to shallow waters or drifted as plankton near the ocean surface. It was here, one presumes, that cilia and flagella first made their appearance. In the course of time substantial deposits of organic debris accumulated on the sea floor, an ever-increasing shower of tiny corpses drifting down from the rich surface waters. With the advent of heterotrophic feeding—the ingestion of plant or animal tissues and remains—this natural resource was soon discovered and exploited. At first the cilia that kept planktonic forms afloat also served adequately to propel bottom-dwellers over their chosen feeding grounds, but cilia are really only suitable for the propulsion of very small animals. A ciliate protozoan 0.1 millimeter in length can easily cover 2 millimeters per second. Nemertine ribbon worms three or four thousand times longer, using the same method of locomotion, can only travel one-tenth this distance in the same time.

Planarian flatworms move primarily by means of cilia, which are abundant on their undersurface, and beat within a layer of secreted mucus. The flattened shape of these creatures not only increases their power base, quite literally, but also improves respiration by reducing the distance that oxygen

Most flatworms are aquatic, but a few, like Bipalium, *have managed to invade the land, where they live in damp environments. They move by a combination of peristalsis and ciliary action.*

must diffuse to the inner tissues. In contrast, terrestrial flatworms strive to minimize their surface area in order to conserve moisture, and consequently are circular in cross-section. In addition, being out of water they weigh more, and hence the cilia must work harder. As a result, terrestrial flatworms, like terrestrial nemertines, are generally slow movers.

The largest animals to rely on ciliary locomotion, apart from ctenophores with their comb plates, are the ribbon worms. Because of their great length — up to 30 meters — their movements are slow, and are mainly used when coiling into crevices and beneath rocks. Both flatworms and ribbon worms are able to improve their locomotory performance by using muscles in addition to their cilia. This they do in three main ways — looping, peristalsis and pedal waves. From an evolutionary standpoint these represent important advances, foreshadowing the massive exploitation of a dynamic, muscle-powered hydrostatic skeletal system, which led initially to an increase in animal size and complexity, and ultimately to the elegant rigid skeletons of both arthropods and vertebrates.

Looping

This is the familiar and very characteristic movement of an inchworm caterpillar. For planarians, with their much simpler musculature, it is, although relatively fast, not a very efficient method of locomotion. For this reason its use is confined to emergencies. A key requirement in looping, as in all "terrestrial" locomotion, is that part of the body remain firmly anchored to the substrate. Lacking any special attachment organs, planarians must rely on the limited friction provided by their mucous secretions. With the rear end anchored, the front part of the body is raised and extended by contraction of the circular muscles acting on the pliable parenchymal cells that fill the body. Once the head end is attached, contraction of the

A fundamental requirement for movement on land is an ability to form strong, temporary anchor points. Leeches have evolved powerful suckers, like this one, at each end of the body to facilitate locomotion, particularly as they crawl over the body of their host.

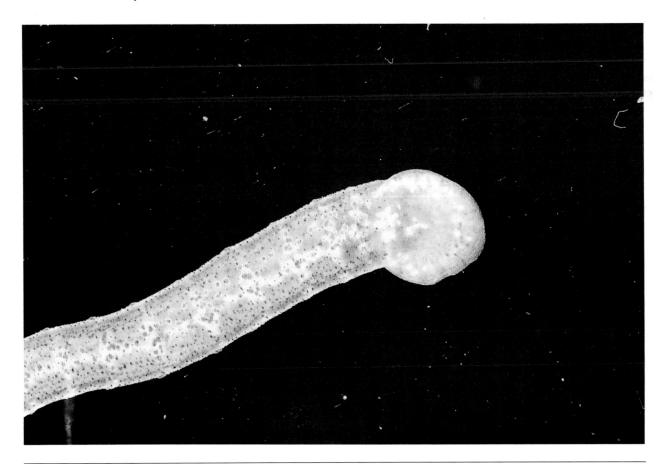

The probascis or ribbon worms are slightly more complex than flatworms and rather larger. Although they use cilia for some movements, their normal mode of progression is very slow peristalsis.

longitudinal muscles draws the rest of the body forward. The parasitic nemertean *Malacobdella,* which lives within the mantel cavity of bivalve mollusks, has taken the first step to improve this method of moving by anchoring the rear end with a sucker that can provide far better attachment than plain mucus. This principle reaches its highest expression in the annelid worms. Here, leeches have not only added very efficient suckers at both ends of the body, but through a more complex interaction of the musculature, significantly increased the length of their stride.

But looping is not the only method of locomotion used by leeches. Using the same musculature and skeletal system, they have also evolved a form of undulatory swimming. Unlike most other swimming animals, leeches oscillate vertically. It is interesting to note that the only other animals that swim in this way, the marine mammals, have also adopted it secondarily.

Leeches have become successful by specializing as predatory bloodsuckers. Yet in doing so they may have entered an evolutionary dead end. By abandoning the typical annelid hydraulic skeleton, based upon fluid-filled coelomic spaces, in favor of more viscous botryoidal tissue, leeches appear to have adopted a specialization that may leave little scope for further advancement. Elsewhere among the invertebrates, the evolution of increasingly sophisticated methods of locomotion seems to be linked to the appearance of more fluid skeletal systems.

Peristalsis

In peristaltic locomotion, waves of contraction sweep the longitudinal muscles of a worm's body from head to tail, producing a series of bulges along its length. These bulges form the anchor points, where force is applied to the substrate. Between the bulges, contraction of the circular muscles extend the body forward. For ribbon worms, with no special organs of attachment and relatively poor coordination between circular and longitudinal muscles, locomotion is slow and inefficient. Even so, these worms can travel some ten times faster by peristalsis than by ciliary action, covering about 100 millimeters in a minute. It should also be remembered that the nemertine body has no coelom, only solid parenchyma like the flatworms. However, it seems likely that the hydraulic system of the proboscis, which extends through the anterior third of the body (where most peristalsis occurs) serves as an alternative fluid skeleton for this slightly more demanding method of locomotion. We shall encounter peristaltic locomotion again in more advanced form when considering the mechanics of burrowing.

Pedal Waves

The third method of locomotion found among the lower worms is that of pedal waves — that is to say, waves of muscular contraction that are confined to the lower surface of the body, which thus functions as a foot. Pedal waves represent a more localized and coordinated interplay of circular and longitudinal muscles than peristalsis, though placing fewer demands on the fluid skeletal system. Each wave consists of lifting a portion of the foot, moving it and replacing it in contact with the ground. The waves of contraction can be either direct (traveling in the direction that the animal is moving) or retrograde (traveling against the direction of movement). The result in either case is a slow but steady movement at a pace faster than ciliary movement can provide. This type of locomotion is particularly characteristic of terrestrial flatworms, who use it in conjunction with their beating cilia. In the genus *Rhynchodemus,* pedal waves are particularly well developed, giving an almost leglike appearance to the swollen temporary attachment points as they sweep along the body from tail to head (Fig. 13). Not only does this provide greater speed and maneuverability but, probably even more important, reduces the amount of mucus required and so

TERRESTRIAL FLATWORM

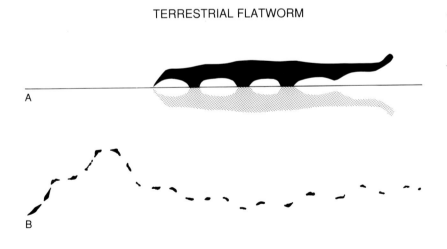

A

B

Fig. 13 Walking in a Terrestrial Flatworm. A) The terrestrial flatworm Rhynchodemus *uses very pronounced pedal waves to produce limb-like attachment points. B) These produce a characteristic pattern of mucus footprints. (After Pantin.)*

helps to conserve both energy and moisture. Instead of leaving a continuous mucous trail, as do snails and slugs, the track of *Rhynchodemus* is a series of discrete footprints. Mucus is deposited only where the swellings are in contact with the ground, partly for adhesion and also to provide an environment in which the cilia can function.

The Molluscan Foot

Chitons and many gastropods such as slugs and snails possess a powerful walking foot. This is a complex organ, which functionally may be likened to the entire planarian body as regards locomotion. It contains an elaborate musculature and a dispersed, blood-filled hydrostatic skeleton, the hemocoel. This consists of a network of sinuses and tiny, interconnected chambers or vesicles scattered throughout the muscular sole of the foot. Although some of the smallest gastropods make use of cilia, most move by waves of muscular contraction sweeping over the foot. If one watches a snail or slug walking up a window pane, the waves of contraction show as dark bands extending across the whole width of the foot and traveling forward over the sole. But this is not so in all species. In a keyhole limpet such as *Diodora*, for example, the waves of contraction are retrograde, pale in color and move independently on each side of the foot.

Direct pedal waves are typically found in the air-breathing pulmonates, even those that have secondarily returned to a life in water. In general the foot of pulmonates resembles that of other gastropods in possessing a thick muscle layer permeated with blood-filled vesicles some 10 microns in diameter beneath the outer epithelial layer. However, there are significant differences in the detailed musculature and in the mucous glands. Pulmonates lack vertically oriented dorsoventral muscles, which are very powerfully developed in some other groups, such as the limpets. Instead, they have a substantial layer of longitudinal muscles, from which oblique fibers run

Mollusks have evolved a characteristic foot, which moves by means of localized contractions and the secretion of mucus with special properties. The foot in snails like this freshwater Planorbis *species has complex musculature which acts on a highly dispersed hydrostatic skeleton.*

down both forward and backward to the sole. Mucus plays an important role in all gastropods, but particularly those that move on land. The precise way in which the mucus functions is not clear. It used to be thought that in pulmonates the anterior mucous gland produced a highly viscous secretion that anchored the foot to the ground, while other glands, distributed over the whole sole, secreted a more fluid kind of mucus serving as a lubricant as the waves of contraction passed. However, recent work suggests a rather different mechanism. Depending on the forces to which it is subjected, the mucus changes its physical properties from an elastic solid to an elastic liquid. In such a system it is clearly necessary that the area of the foot beneath which the mucus is yielding be less than that serving for anchorage.

It was noted above that in pulmonates the direct waves of contraction appear dark. This is so because in the parts of the sole anchored to the substrate, the muscles are relaxed. In the dark band, contractions of the posterior oblique muscle fibers compress the local vesicles and lift the sole very slightly away from the surface. This is accompanied by secretions of mucous, which fill the resulting space, allowing the contracted region to move forward. In this way, each section of the foot is drawn forward and reattached.

In a limpet, for example *Patella,* the process is different. The muscle layer consists primarily of dorsoventral fibers, with some transverse ones. The longitudinal fibers typical of pulmonates are largely absent and play no part in locomotion. At rest, *Patella* is attached to its rock by the viscous adhesion of its mucus. When disturbed, either by intending predators or by wave action, contraction of the dorsoventral muscles lifts part of the sole to create a vacuum beneath. When *Patella* leaves its home site to browse, locomotion results from the interplay of the transverse and dorsoventral muscles on the fluid within the hemocoelic vesicles. Contraction of the muscles lifts the sole some 0.2 millimeter off the substrate, and causes the vesicles to elongate. As the wave passes and the muscles relax, internal pressure restores the vesicles to their normal spherical condition. This causes the sole to shorten once again as it comes back into contact with the substrate and restores adhesion. Because the retrograde wave is one of relaxation, it appears paler than the adjacent contracted areas of the foot, which serve as anchor points. In normal movement, slightly more than half the sole is lifted at any instant.

Gaits

We have already noted that in *Diodora* the foot is divided longitudinally into two independent functional units. The same kind of division is found in many other mollusks, taking a diversity of forms and resulting in a variety of distinct gaits — analogous to walking, trotting or cantering in a quadruped. Although it is generally unclear precisely what benefits such slow-moving creatures might gain from the evolution of different gaits, they appear to give greater agility. In the case of limpets, for example, it allows an individual to rotate on its axis while remaining in one spot. By this means the shell and rock can be ground together to make a perfect fit, so lessening the chance of accidental displacement. In the active and agile abalone *Haliotis,* the two halves of the foot are each divided into three regions, one attached and two moving or vice versa. Thus at any instant there is a triangular system of support in effect, reminiscent of that we shall encounter again when considering the six-legged walking of insects.

In the land winkle, *Pomatias,* the trend toward specialization has been carried even further. Here, while one half of the foot remains anchored, the other is lifted free of the ground and moved forward. This results in a most curious shuffling form of locomotion. Other gastropods, including snails belonging to the genus *Helix* and sea hares of the genus *Aplysia,* make use, on occasion, of a fast, loping gallop quite unlike their normal progression. This consists of large retrograde waves, which are superimposed on the typical slower, smaller, direct waves.

The most extreme development of a loping gait is to be found in the large West Indian queen conch, *Strombus.* Using its massive shell as an anchor, the foot is fully extended, until the tough, bladelike operculum, the cover that seals the shell closed, can be driven into the sand. With the foot thus firmly anchored, the longitudinal muscles contract and the shell lurches forward for the cycle to be repeated.

Burrowing

This quill worm (Hyalinoecia) *is a typical burrowing polychaete, which secretes a tube of thick, parchmentlike material. Although the worm can withdraw into the protection of its tube when danger threatens, it normally lives with only the hind part of its body there, crawling about on the adjacent sea floor with its parapodia.*

At first, life for the creeping denizens of the ocean floor was easy as they devoured the accumulated organic debris of past ages. But predators, together with competition for space and resources, gradually increased pressure in favor of evasive and defensive specializations. The story of evolution is the record of the way living organisms have responded to these pressures. Animals feeding on the soft, muddy bottom deposits had an easy life at first, but all good things must end. Natural selection was about to force another advance in the battle for survival. One way to escape the hazards of life on the sea floor is to burrow. But burrowing proves to be a

highly specialized activity, and we~~~~~~~~~~~~~~~ut appropriate
adaptations. In fact, three separate types ~~~~~~~~~~re required for a
successful burrowing existence. First, a means ~~~~~~rating the surface;
second, a means of excavating; and third, an appropriate mechanism for
traveling through the completed burrow. As we shall see, special problems
arise when trying to burrow in hard substrates such as wood or rock, but
even these have been successfully overcome by a number of phyla, most
notably by certain bivalve mollusks.

Penetration

Protection from predators and the promise of vast food reserves can only
be realized once the problem of penetration is solved. In this regard, Newton's third law of motion presents enormous difficulties, because small
organisms find it hard enough anchoring themselves for horizontal movement, let alone digging vertically downward. For mollusks, the weight of
their shell provides some reaction to probing movements of the foot during
penetration, but for soft-bodied creatures whose density is close to that of
seawater, the problem is often insuperable. Some phyla, such as the wormlike priapulids and sipunculids, have never mastered the art as adults. They
depend on the ability of their tiny mobile larvae to locate a suitable settling
place and establish a beachhead between the sand grains within which
metamorphosis and growth can occur. Once established in this way, priapulids and sipunculids seldom, if ever, voluntarily leave their burrows.
Lacking any means of anchoring their body, a return to the mud is only
possible if fate presents a suitable crevice into which they can insinuate
themselves.

In contrast, those species that have evolved the ability to burrow down
from the surface not only enhance their chances of survival if accidentally
displaced, but are able to exploit a much greater range of habitats. The
benefits of such flexibility are best exemplified by the West Indian surf
clam, *Donax,* which is able to migrate up and down the beach with the
changing tides to exploit the rich pickings of the ever-shifting surf. Only
three seconds is needed for *Donax* to anchor itself sufficiently against displacement by wave action. Able to dig in at 4 millimeters per second,
Donax can penetrate sand about 10 times faster than its more conventional
relatives. Such athletic prowess, of course, necessitates an abnormally high
expenditure of energy. However, for these active and successful little
bivalves, the sums work out satisfactorily because they are able to feed
continuously, regardless of the state of the tide, in a rich habitat and virtually without competition.

But the ability to burrow is not confined to higher invertebrates. Suprisingly, some sea anemones, with a level of organization simpler than even
the flatworms, have become good burrowers. This is remarkable, for burrowing in sand would seem to require considerable strength, and yet the
hydraulic skeleton of an anemone such as *Peachia* can generate only the
faintest pressure. The secret of its success lies not in strength, however, but
in persistence. *Peachia* is distinguished from other anemones by having the
basal end of its column modified into a digging organ in place of the usual
adhesive disc. Lying on the surface, the tip of the column is turned downward, and a slow sequence of peristaltic movements begins. These cause the
knoblike digging organ, or physa, first to be turned inside out with a
corresponding displacement of sand, and then inverted by the contraction
of longitudinal muscle fibers in the trunk. As sand is displaced during
inversion, the trunk is slowly drawn into the sand. Each cycle of eversion
and inversion lasts about 90 seconds, so that an hour or more is required
for the trunk to be buried. The initial stages are particularly slow because,
for a time, the trunk cannot be anchored to provide the physa with a firm
base against which to push.

A mechanically similar mechanism is found in some of the burrowing annelids such as the lugworm, *Arenicola*, which has an eversible proboscis (Fig. 14). Inflatable flanges equipped with spines (chaetae) just behind the proboscis help anchor the worm in these early stages of digging. Because there is little to press against, the proboscis of *Arenicola*, like the physa of *Peachia*, is not a powerful bulldozer but a delicate and persistent tool, gradually displacing individual sand grains. This slow and gentle approach is common to many burrowing forms. Acorn worms, such as *Saccoglossus*, employ an eversible proboscis like *Arenicola*, but the marine gastropod *Natica*, which has an exceptionally long and mobile foot, uses ciliary action in the initial stages of burrowing.

It seems highly likely that for many creatures, the thixotropic properties of sand and mud play an important role in burrowing. Thixotropy is a property of small particles in suspension to change from a solid to a fluid state in response to suitable mechanical stimulation. Although it is difficult, if not impossible, to force one's hand straight into damp sand on the beach, gentle vibration enables it to sink in easily. It is the thixotropic properties of quicksand that make it dangerous.

Excavation

Once the initial penetration is complete, burrowing proper can proceed. This has two phases, and is well exemplified by *Arenicola*. First, the embedded anterior segments shorten and dilate to provide an anchor that will allow the following segments to be drawn in. Then the configuration changes to provide a penetration anchor. In this, the embedded segments dilate toward the rear to form a series of flanges that, together with protruding chaetae, very effectively prevent any backward movement. While anchored in this way, the proboscis is everted and then withdrawn. This sucks water into the end of the burrow, softening the adjacent sand, which yields to the next forward thrust. Once established, *Arenicola* maintains the shape of its burrow by lining it with a layer of mucus. It also appears to maintain the burrow and increase its diameter by forcefully expanding the body. This action is accompanied by brief pulses of high pressure within the fluid coelomic skeleton. *Arenicola* is highly adapted for burrowing, and few other polychaetes possess a comparable proboscis. Most species have to rely on their crown of tentacles for digging.

In bivalve mollusks, the foot equates functionally to the anterior segments of *Arenicola*, alternately thrusting and dilating. However, anchoring the body during penetration is accomplished by relaxation of the adductor muscles that close the shell. This allows the shell to open under the influence of the elastic hinge ligament, so wedging the shell against the walls of the burrow. Burrowing bivalves exhibit a variety of special adaptations that reflect their particular life-style. For example, some clams possess enormously elongated siphons that enable them to maintain feeding currents while deeply embedded. Species that specialize in rapid digging, such as the razor clams *Ensis* and *Solen*, have long, parallel-sided shells and an extremely mobile foot.

Life in a Burrow

Many burrowing animals excavate a semipermanent home in which they live and move about. This is done using peristaltic movements similar to those employed for digging, and may be either direct or retrograde. It will be recalled that in the case of retrograde waves, anchor points are formed by shortening of the longitudinal muscles. Within a burrow, thickening of the anchor points provides secure attachment (Fig. 15). Likewise, extension of the body between the anchor points, brought about by contractions of the circular muscles, releases the attachment to the burrow wall. This happens because each segment functions as a separate unit with its own, independent hydrostatic skeleton acting in concert with antagonistic sets of

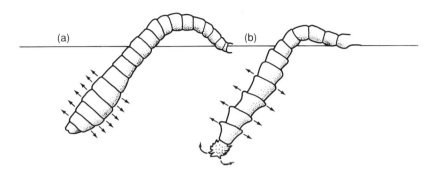

Fig. 14 Burrowing Lugworm. The lugworm (Arenicola) *uses two patterns of muscular activity to burrow. First (A) the anterior segments dilate to form a terminal anchor. This is accomplished by contractions of longitudinal muscles in the segments behind the anchor. Next (B) the longitudinal muscles relax and the circular muscles in the same segments contract. This converts the buried segments into a series of enormously powerful flanged anchors. Thus stabilized, the pharynx can be forcibly everted to extend the burrow.*

circular and longitudinal muscles. These are the movements typical of earthworms.

In free-living animals that progress by the use of direct waves, it is the contracted segments that move and the extended segments that provide anchor points. Such a mechanism is clearly unworkable within the confines of a burrow (Fig. 16). Nevertheless, *Arenicola* and many other burrowing polychaetes are found to make use of direct waves. How can this be so? The answer is to be found in the nature of their hydrostatic skeleton. Unlike the earthworm, in which each segment is isolated by septa from its neighbors, the coelom of the trunk region in worms like *Arenicola* is continuous from segment to segment. This allows maximum girth and maximum extension to occur together in the same segments through simultaneous relaxation of both circular and longitudinal muscles. The system is made more efficient by the presence in some species of a complex network of elastic collagen fibers beneath the epidermis.

Amid the more specialized tube-dwelling (tubicolous) polychaetes, such as the lugworms, the parapodia are much reduced, a trend ultimately leading to the oligochaetes such as the earthworm, in which parapodia are completely lacking. However, some tubicolous polychaetes move within their burrows by walking like their free-living relatives. This type of locomotion, the forerunner of typical arthropod movements, will be considered in more detail in a later section.

Boring

Amid the many extraordinary achievements of invertebrates, one of the most remarkable is an ability to burrow into hard wood and rock. This talent is possessed by a variety of phyla, including sponges, flatworms, annelids and mollusks. Although aided by acids and other chemical secretions in some cases, the principal method of penetration appears to be slow, persistent mechanical abrasion. This is particularly so among bivalve mollusks. It is within this class of mollusks that are found the most proficient exponents of the art of rock boring, which has arisen independently several times in unrelated genera.

Boring among such bivalves is the outcome of persistent abrasion brought about through small, repetitive movements of the shell, which is often modified in shape and provided with ridges and other special sculpturing. Although high internal pressures, like those encountered in specialized sand burrowers such as *Ensis,* are not typical of rock borers, nevertheless hydraulic adaptations are found in many species. These involve modifications to the mantle cavity which generate pressure that forces the valves of the shell apart more strongly than could be achieved by an unaided hinge ligament. The success of a bivalve shell as an anchor depends on the force it can exert on the burrow wall. Thus the need for more powerful anchorage can be met only by increasing the outward thrust of the valves, once the shape of the shell has been modified to provide the maximum area of contact.

Another adaptation widespread among burrowing bivalves is a change in the orientation of the axis about which the shell opens. This is normally aligned front to back, but secondarily becomes top to bottom, and is associated with a reduction in the hinge ligament. This modification allows the siphons to remain extended when the anterior margins of the shell are drawn together in digging. Even more important, it helps to enhance the hydraulic function of the mantle cavity. Such a system is well developed in the clam *Mya* and in the piddocks, such as *Pholas* and *Zirphaea,* which are perhaps the most adept rock borers. The rocking movements of the shell, caused by antagonism between the siphon muscles and the adductors, are supplemented by rotary movements, brought about by use of the foot. However, not all burrowers make use of the foot in this way.

Mussels and their relatives anchor themselves to the substrate by means of byssal threads. These tough, horny strands are initially secreted as a fluid by glands in the foot, but subsequently harden through changes in chemical structure. In burrowing forms such as *Botula* it is movements of the shell resulting from contractions of the byssal retractor muscles that lead to abrasion of the surrounding rock.

The most specialized boring mollusk is the shipworm *Teredo,* which inhabits saturated wood and is a major foe of piers, pilings and ships. The greater part of *Teredo's* wormlike body consists of the mantle, which is drawn out to form an elongated siphon. The shell is greatly reduced, and consists of two small valves that are hinged from top to bottom. Thus they can be opened and closed by the antagonistic action of the anterior and posterior adductor muscles. The anterior edge of the valves is modified into a cutting organ consisting of sharply ridged lobes, which must be continually replaced as they wear away. These efficient cutters rasp off wood as contraction of the posterior adductor forces them apart. Between each scrape, the foot emerges and adheres to the end of the tunnel. It then rotates the shell slightly, and draws the cutting edges forward for the next cycle. Thus *Teredo* is distinguished from other burrowers by relying entirely on direct muscular action, with no intermediate hydraulic mechanism.

The tunnels made by shipworms generally run with the grain, but seldom emerge on the surface. This is believed to be due to the animal's instinctive avoidance of the surface wood, which tends to be soft and rotted through the action of fungi and bacteria. For most organisms that bore into hard substrates, the evolutionary pressures responsible for this behav-

Figs. 15 and 16   Passage of Direct and Retrograde Waves. These diagrams illustrate the differences between direct and retrograde waves along the bodies of moving worms. Each segment is identified by a letter (A–H). The state of contraction of each segment is out of phase with its neighbor by one-eighth of a cycle, and is represented by Roman numerals. In a direct wave it is the segments that are most elongated that are stationary and provide the attachment points. Conversely, in a retrograde wave it is the segments shortest and broadest that provide this service.

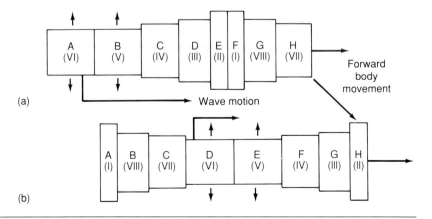

PASSAGE OF A DIRECT WAVE

ior appear to be related to protection. However, this is not so in the case of *Teredo*, which consumes as food the sawdust it manufactures. This is made possible by the presence of symbiotic bacteria in the gut, which serve a similar function to the specialized cellulose-digesting protists in the gut of termites.

Tube Feet and the Echinoderms

From simple beginnings among the lower worms, we have been following the gradual sophistication of locomotory mechanisms as we progress up the evolutionary ladder through the mollusks and annelids. We have seen how the success of certain groups is associated with their exploitation of specialized modes of locomotion—for example, jet propulsion and buoyancy control in cephalopods, or looping and attachment suckers in leeches. Nowhere in the animal kingdom is this principle more clearly demonstrated than among the echinoderms.

As with so many other major groups, echinoderm origins lie concealed in the uncharted mists of the Precambrian era. All living echinoderm classes are united through the possession of tube feet. These remarkable organs exhibit a wide variety of form and function. They can be tactile, respiratory, locomotory, chemosensory, feeding, burrow-building and more. It seems probable that they arose initially as respiratory organs, but it is their role in locomotion that interests us here. The tube feet are essentially saclike outpushings from a very complex coelomic system that protrude through the skeletal plates of the body wall. The typical tube foot of a starfish or sea urchin consists of an extensible cylinder and terminal sucker. The side walls contain longitudinal muscles and an elastic sheath of collagenous connective tissue within a flexible cuticular layer. Inside the body, beneath the skeletal plates, the tube foot is connected to an ampulla, a muscular sac that is joined to a water-vascular system that runs throughout the body. Each ampulla and its associated tube foot functions as an independent hydraulic system. Contractions of the ampullate muscles force water into the tube foot, which extends accordingly. Movement of the foot is controlled by differential contraction of the longitudinal muscles as well as by a ring of postural muscles at the base that attach to the ring of skeletal plates.

Pressure changes within the coelom of echinoderms are slight, and it is the muscles of the tube foot that provide the system with strength. Even though the power of a single tube foot is slight, because of their great number the total effect can be substantial, as evidenced by the ability of starfish to separate the shells of bivalve mollusks. This is achieved not so much by brute force, but by steady, unremitting effort over an extended period. The powerful cumulative effect of the tube feet, combined with a heavily armored skeleton, enables some echinoderms to colonize exposed

PASSAGE OF A RETROGRADE WAVE

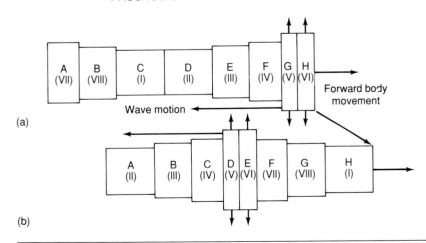

The peacock worm (Sabella) is another, more specialized tube dweller. The only part to protrude from the tube, which is constructed of sand grains held together by mucus, is a magnificent crown of tentacles used to filter food particles from the surrounding water.

shores where they are able to withstand the full force of the surf as they cling to the rocks.

The tube feet of echinoderms are perhaps the ultimate manifestation of the use of hydraulic systems for locomotion. Although clearly lacking any potential for speed and agility, tube feet have made a substantial contribution to the evolutionary success of the phylum. This has only been possible through the parallel development of an appropriate nervous system that coordinates and integrates the operation of the entire army of tube feet.

Thus we find among aquatic animals a wide diversity of locomotory mechanisms, and within each category a wealth of special refinements and adaptations. It was from such a repository of animal form and function that the invasion of the land was launched. Millions of years of selection and adaptation were about to be thrown back into the evolutionary melting pot for this major advance to occur. Haltingly, tenuously, the struggle for mastery of the land began with the demands of a totally different set of design parameters. Just how the forces of selection adjusted to this challenge we are about to see.

Mussels (Mytilus) *attach themselves to rocks by means of special byssal threads. These are laid down, one by one as shown here, as secretions produced by the very mobile foot. They soon harden into a network of tough guy-ropes.*

The forest of tube feet protruding from this sea urchin (Echinus) *is typical of echinoderms. They are used for feeding, cleaning and locomotion, and although individually quite weak, achieve strength through numbers.*

Polychaetes and the Origins of Crawling

In order to understand the next great advance in locomotion, the invasion of the land by arthropods during the Silurian epoch, some 450 million years ago, we have to start many millions of years earlier and consider the evolution of the arthropods themselves. In particular we must look at the way in which limbs might have arisen, and this means returning to the annelid worms, particularly the polychaetes.

The polychaetes that have attracted us thus far have been largely burrowing forms. However, there are a large number of free-living or errant, species. Of course, modern polychaetes are the culmination of many millions of years of selection and specialization, and cannot, therefore, provide

The anemone Calliactis *attaches itself to shells inhabited by hermit crabs* (Eupagurus). *This is a classic example of commensalism. The anemone feeds on food particles torn off by the crab, which in turn receives protection from predators by the anemone's stinging tentacles.*

any direct arthropod ancestors. Nevertheless, we can draw some inferences about ancestral polychaete stock back in the frustrating obscurity of the late Precambrian.

The body of a typical errant polychaete like *Nereis* is built up of serially repeated segments. Each is traversed by blood vessels, the gut and a ventral nerve cord. In addition, each contains dorsal and ventral longitudinal muscles, divided into a left and right bundle. Most segments possess a pair of hollow lateral extensions, the parapodia, which are also well equipped with muscles. The most important of these are the extrinsic muscles that run obliquely fore and aft to the midline of the body on both top and bottom. These muscles act against the pressure within the coelom, which extends into the parapodia, and is separate within each segment. In addition, each parapodium has its own intrinsic muscles, which extend and withdraw the bundles of conspicuous spines, or chaetae. Another significant feature of the *Nereis* body plan is the weak and incomplete arrangement of circular muscles in the body wall of each segment. Thus, *Nereis* cannot use the kind of peristaltic movements that we have previously seen in the lugworm and earthworm. Instead, independent, out-of-phase contractions of the left and right longitudinal muscles throw the body into a series of undulations. Many errant polychaetes can swim using these undulatory movements in combination with paddling movements of the parapodia, but it appears that this type of locomotion has evolved as a secondary means of progression from less violent crawling movements.

Slow crawling depends solely on movements of the parapodia, which function simply as a series of projecting levers. Each parapodium, which is half a cycle out of phase with its partner on the opposite side, is raised and thrust forward just ahead of its neighbor in front. Hence, the waves of movement are direct, rather than retrograde. Once in contact with the substrate and anchored with the help of extended chaetae, the posterior oblique muscles contract and pull the parapodium backward. This happens repeatedly along the body, with numerous anchor points serving to propel the worm forward in a uniform continual motion. The parapodia are small in relation to the total body mass. Consequently little energy is lost to inertia through repeatedly stopping and starting. This is in contrast to peristaltic locomotion, in which the whole body is alternately accelerated and decelerated.

Simple movements of the parapodia provide polychaetes such as *Nereis* with an economical way of moving over the sea bottom, but they are not well suited to escaping from predators. For this, the whole body is convulsed into a series of sinusoidal waves through contraction of the powerful longitudinal muscles. In this situation, an anchored parapodium is at the crest of a wave and its partner, in the opposite trough, is moving forward. Thus the anchor points are associated with both maximum extension and maximum relaxation of the driving muscles. It will be recalled from the consideration of *Helix* that this is the condition in which direct rather than retrograde waves are operative.

Introduction to the Arthropods

The arthropods are united by the possession of a tough cuticle that provides armor-plated protection against the outside world and a rigid skeleton against which muscles of the body can act. This represents a major functional departure from the hydrostatic skeletons of other invertebrates. The benefits of this radical new design cannot be overstated, and are clearly shown by the fact that arthropods substantially outnumber the whole of the rest of the animal kingdom combined.

The parapodia of this rag worm (Nereis) are used both for walking over rocks or sand and as paddles when swimming. Although the parapodia of modern polychaetes are too specialized to have given rise to the limbs of arthropods, they nevertheless provide clues to the way in which arthropods may have evolved.

Because arthropods are so distinctive, it was natural to assume that they all arose from a common ancestral stock and are thus related. However, detailed analysis, particularly of their functional morphology, makes it clear that the processes of arthropodization evolved independently on several occasions among serially segmented coelomate worms. For this reason we ought properly to regard the major branches of the arthropods as separate, unrelated phyla. The apparent similarities between these groups are misleading. Given a tendency toward toughening of the cuticle among worm-like ancestors, comparable mechanical demands inevitably resulted in the convergent evolution of solutions that were superficially very similar.

Segmental appendages arose in these ancestral forms, as they have done later in polychaetes, for a variety of purposes that include filter-feeding, respiration, prey capture, sensory perception and, of course, locomotion. This diversity of function is well illustrated today by the Crustacea, whose limbs show an extraordinary range of form.

The acquisition of arthropod characteristics by predators of soft-bodied bottom dwellers bestowed substantial advantages, and evolution repeated itself several times with relatively minor variations. In addition to the three

Horseshoe crabs (Limulus) *are not crustaceans but chelicerates — relatives of spiders and scorpions. They are living fossils that have persisted almost unchanged for about 400 million years and are important in understanding early arthropods. They wander over the sea floor digging for burrowing worms, mollusks and crustaceans.*

main arthropod groups alive today—the crustaceans, the chelicerates and the myriapod/insect line (Uniramia)—there have been a number of other, important evolutionary experiments. For more than 250 million years, from the Cambrian to the Permian, the dominant predators of the seas were the trilobites, for whose total extinction there is still no clear explanation. Sharing this dubious honor with the trilobites were the eurypterids, another Paleozoic success story, which today has just one surviving representative, the horseshoe crab *Limulus*.

Less successful in the race to leave descendants, but no less remarkable, are other unique experiments in arthropodization: mysterious creatures with sonorous names like *Cheloniellon*, *Marrella* and *Arthropleura*. The most extraordinary of these extinct arthropods, and also the best preserved, lived some 530 million years ago in a single community of over 140 species in western Canada, and are preserved in the Burgess Shales. Some, such as *Aysheaia*, appear to be aquatic relatives of the velvet worms (Onychophora), but others, such as the stilt-legged *Hallucigenia*, belong to lineages long since extinct and with no recognizable affinities.

Whip scorpions (Mastigoproctus) are arachnid relatives of spiders and scorpions. They use their front pair of legs as long antennalike sensory organs, walking like insects on the remaining three pairs.

THE ORIGIN OF LEGS

Walking must have evolved independently many times in the course of evolution. The sea mouse *Aphrodite* is one such example. In this somewhat aberrant polychaete the longitudinal muscles and the oblique extrinsic muscles running to the parapodia are both reduced. However, the internal muscle of the parapodia are strongly developed, allowing the animal to move with powerful and rapid retrograde steps and without wasteful lateral

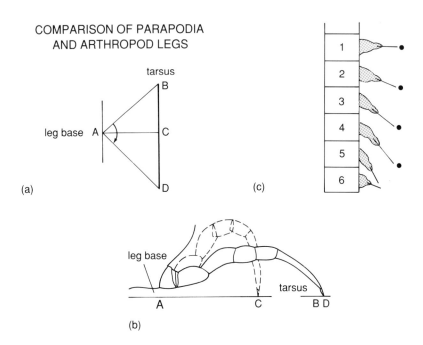

Fig. 17 Limb Movement in Polychaetes
and Arthropods. The jointed leg of an
arthropod starts its stride extended, flexes
in mid-stride and then extends again.
This is shown in plan view and
horizontally in (a) and (b). The
parapodium of a polychaete (c) cannot
bend or flex. Instead, the terminal spines
(aciculae) are extruded and withdrawn to
maintain static contact with the ground
(marked as black dots) as the animal
moves forward. (After Manton.)

body movements. But modern annelids are the product of millions of years of natural selection, and they have acquired many specializations since their differentiation in the late Precambrian. The parapodia of polychaetes are just one such specialized acquisition, and there is no way in which arthropod limbs could have arisen from such a structure. The musculature, the hydrostatic skeleton and the basic mechanical principles of operation are all wrong. Consider a parapodium anchored to the substrate at its tip by sharp chaetae. As the body moves forward and the parapodium swings back, the chaetae must be increasingly extruded to prevent slipping (Fig. 17). This done by intrinsic muscles, but the amount of such extension, and hence the time during which force can be applied by the parapodium, is limited. In a typical arthropod, the distinctive geometry of the leg allows much longer strides.

Peripatus

A clue as to how arthropod limbs arose is provided by *Peripatus*, a member of the Onychophora, or velvet worms. These curious tropical creatures live in rotting logs, and possess a mosaic of characters typical of both annelids and arthropods. In particular, their dumpy legs, totally distinct in structure and function from parapodia, suggest a route whereby arthropod limbs could have arisen. Again, the fossil record is sadly incomplete, but we do possess *Aysheaia*, a remarkable mid-Cambrian marine creature, whose well-preserved limbs closely resemble those of *Peripatus*.

Because of its unique blend of characters, it is tempting to view *Peripatus* as a "missing link," an archetypal primitive arthropod precursor. This is deceptive and misleading. *Peripatus*, though providing clues to its ancestry, is nevertheless a specialized and highly adapted creature, many of whose morphological features can be directly attributed to its life-style. Its ability to squeeze through extremely small crevices, for example, precludes the use of armor plating and conventional limbs. Even so, the velvet worms use their limbs in a way that unites them with the myriapods and insects and sets them apart from other arthropods. This distinguishing characteristic is the use of different gaits. We shall examine this subject in more detail later on, but in essence these animals are able to alter the timing and sequence of leg movements in order to attain higher speeds. Such coordination is seemingly byond the capabilities of both crustaceans and chelicerates.

Opposite:
*This paper wasp (Polistes) illustrates very
clearly the typical organization of the
insect body. The head bears both simple
and compound eyes. The thorax has two
pairs of wings in addition to the six legs,
and the segmented abdomen lacks any
appendages.*

Jointed Legs

At first sight the arthropods present a bewildering array of body forms and an incomprehensible profusion of limbs. Careful analysis, however, reveals distinct patterns, and it is now possible to correlate many anatomical features with quite specific life-styles. Thus millipedes, which live in rotting vegetable debris and have to force their way slowly through compacted compost deposits, have a specially strengthened body and numerous short, powerful legs. In contrast, fast-moving predatory centipedes have fewer, longer legs with specialized joints that increase the angle through which the legs swing in each stride. What confuses seemingly straightforward principles of this sort are secondary adaptations, such as the evolution of a fast-moving, scavenging or predatory life-style by certain millipedes and the adoption of a burrowing life-style by elongate geophilomorph centipedes. The superimposition of apparently conflicting adaptations in this way can obscure the path of evolution, and make interpretation a fascinating, if complex, piece of detective work.

Arthropod limbs are generally cylindrical and packed with muscles. These are usually antagonistic pairs, which control the angle of the joints with which they are associated. The principal locomotory forces generated by an arthropod limb are associated with a forward-and-back swing (known technically as promotor and remotor) which is generated by extrinsic muscles within the body acting on the coxa, the leg segment closest to the body. The articulation of the coxa and other segments is a matter of great significance, and the design of these joints exhibits a multitude of special adaptations. First, there is the nature of the hinge. This can range from a simple softening of the cuticle that allows slight bending, through to a pair of condyles (articulation points) providing large, powerful flexure

This velvet worm (Macroperipatus) *shares features with both insects and annelids, and in consequence has attracted a great deal of evolutionary attention. Although highly adapted to their specialized habitat in rotting logs, velvet worms provide valuable clues to the origins of both insects and myriapods.*

This burrowing geophilomorph centipede (Haplophilus) has a long, very flexible body with short legs. In some ways its general form is more reminiscent of millipedes than of its surface-dwelling, fast-moving relatives.

about a fixed axis. At other joints, a single dorsal condyle allows movement in two planes, but this necessitates the evolution of special devices to prevent the two elements from becoming separated. In addition, there are adaptations to permit greater and greater angles of flexure through allowing elements opposite the hinge to telescope into one another. In some millipedes this is even accompanied by fluid-filled joints providing a smooth, frictionless action that has been likened to the use of synovial fluid in the joints of mammals.

When it was believed that all arthropods descended from a single, common ancestor, it was unquestionably accepted that all limbs were derived from one fundamental pattern. Bitter, fruitless arguments raged over the true relationships of the various elements and which might be homologous. The limbs of the different arthropod groups are diverse and variable. Such a rigid one-to-one correspondence of the parts is far from obvious. All manner of ingenious explanations and special pleadings were invoked in order to try to make the observed facts fit the procrustean bed of accepted dogma. This proved to be wholly misguided and unnecessary, for it is now clear that the number of segments into which a limb is divided is directly linked to the manner in which the limb is required to perform, and is not a cryptic component of some great plan of preordained cosmic homologies.

The number of legs, too, is something of an adaptive feature. Burrowing forms require the combined strength of numerous short limbs, each remaining in contact with the ground for as long as possible in each stride. In fast running, however, long, rapidly moving legs are advantageous, but prone to becoming entangled with one another. Among myriapods, particularly centipedes, there is a clear reduction in number of limbs associated with increasing speed of locomotion. In fact, there is an overall tendency for the number of walking legs to be reduced to six, as we typically find in insects. Even among arachnids, which usually possess four pairs of walking legs, it is quite usual to find the front pair being used not for locomotion but for prey capture, for courtship display or as sensory structures. The most extraordinary legs of all are those of the sea spiders (Pycnogonida), marine arthropods of uncertain affinities. Not only does the number of walking legs vary from four to six pairs in different genera, but the body is so reduced that it serves as little more than a site for articulation of the limbs. The internal organs of the body have been displaced into the legs, and females may even be found with eggs developing there.

The legs of dogs and horses project straight down, supporting their owner's weight high above the ground. Among arthropods, the legs are

The legs of higher vertebrates, such as this topi (Damalaiscus), project straight down, supporting the body high above the ground. The resulting reduction in stability is made up by greater maneuverability and speed.

Fig. 18 Arthropod Limbs. With the exception of isopods such as sow bugs, most arthropods gain stability by hanging suspended from their legs, as shown in these diagrams of a typical insect, spider and crustacean.

LEG ARRANGEMENTS IN VARIOUS ARTHROPODS

Earwig

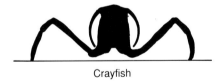

Crayfish

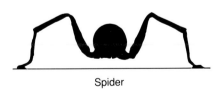

Spider

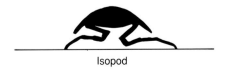

Isopod

generally inserted quite differently. If an insect were to perch on high legs like those of a dog, the slightest breeze would blow it over. For this reason, arthropods do not stand on their legs, but hang suspended from them (Fig. 18), often keeping the body very close to the ground as a result. One consequence of this arrangement is that it makes the knee joint and the coxa of prime importance, and it is at these joints, more than at any other, that the greatest range of adaptive specializations is to be found. These include, for example, modifications that enable rapid runners to increase the length of their stride.

An exception to this rule is found in the arachnids. In these animals, which include spiders and scorpions, the coxa is firmly fused to the body and projects out sideways. The source of mobility in the legs of arachnids is the following segment away from the body, the trochanter, and in consequence the actual orientation of the hinge at this and subsequent joints varies from group to group depending on how the limbs are used.

The limbs of crustaceans differ markedly from those of other arthropods, in both their structure and their manner of use. Crustacean limbs, like those of the merostomes and trilobites, are biramous, that is to say, divided into two branches. They come in a bewildering range of different forms, the great majority being adapted for feeding, respiration and swimming.

Although it seems probable that the Crustacea arose from crawling bottom-dwellers, their limbs are never markedly well adapted for walking, despite their evident plasticity in other applications. This is presumably related to their essentially aquatic habit, in which swimming bypasses the need for the evolution of sophisticated joints and hinge mechanisms for increased speed and agility in running. Nevertheless, they do possess some unexpected specializations.

We have previously commented upon the use of the antennae for swimming, and there exist some other, even stranger changes in function. The

Harvestmen, relatives of spiders, have very long, slender legs that nevertheless support the tiny, globular body in a very stable manner by allowing it to hang suspended. The legs end in numerous small segments that allow them to be twisted around grass stems and other vegetation.

No legs in the animal kingdom are more extraordinary than those of the sea spiders (Pycnogonida). The basal segments contain vital organs displaced from the greatly reduced body.

fish louse (*Argulus*) lives partly as an external parasite attached to the body of its host. The first maxillae of *Argulus,* which in most crustaceans are an element of the mouthparts, have become modified as suction pads to allow the creature to adhere to its host. These structures somewhat resemble the tube feet of echinoderms in appearance, and are used in a curious kind of shuffling walk as *Argulus* moves over its host's scales, pausing from time to time to insert its proboscis and feed.

Crabs, and particularly wood lice or sow bugs (Isopoda), are among the few crustaceans to live successfully on land, and it is interesting to note that the latter have evolved a unique way of folding their legs beneath the body (Fig. 19). This gives them the mechanical advantages of relatively long legs, and at the same time keeps them well protected.

Arthropod Body Plan

The overall shape of an arthropod is often a clear indicator of its mode of life. Species that live beneath bark, for instance, tend to be flattened, and the same is true of polydesmid millipedes that have to force their way between compacted layers of rotting leaves. There is, however, a more fundamental way in which the arthropod body can be affected by function. This is the phenomenon of tagmosis, or the division of the body into functional elements formed by the fusion of the basic segments. The various arthropod groups all arose from segmented wormlike ancestors. Evi-

The fish louse (Argulus) *is a crustacean highly adapted to creep over the scales of fish, on which it feeds by sucking out blood. It can swim free for short distances by beating its legs, but normally it holds tight by means of its large suckers and associated hooks and spines.*

dence of this is still to be found in embryological development and in aspects of detailed internal anatomy. However, this primitive segmental arrangement is largely obscured by a secondary regrouping of the ancestral segments into functional units, technically called tagmata but often also confusingly referred to as "segments." This can be readily illustrated by reference to insects.

The body of a typical insect is divided into three functional units, or tagmata: the head, thorax and abdomen. The head consists of a fusion of six embryonic segments, the second, fourth and sixth bearing appendages (antennae and mouthparts). The thorax comprises three segments, each bearing a pair of legs, the second and third also bearing wings. The abdomen consists of eleven segments, usually clearly demarcated and bearing breathing pores or tracheae. It should be emphasized that the presence of tergites and sternites, toughened dorsal and ventral plates, do not necessarily indicate the presence of true segments, but may be a manifestation of later tagmosis.

Tagmosis, which has become stable among insects and arachnids, is variable within the myriapods. The diplopods or millipedes are basically designed for burrowing, using a large number of short, powerful legs. The body of a typical julid millipede is composed of short, rigid circular units, each bearing two pairs of legs. These are tagmata, each composed of two fused embryological segments. This arrangement allows adjacent elements to telescope into one another, which not only imparts strength to resist bending, but also allows the animal to coil up into a defensive spiral when necessary in order to protect both head and legs.

Although spiders lack extensor muscles at the knee joint this does not interfere with walking. Male tarantulas (Aphonopelma) *cover considerable distances searching for a mate during their brief adult life.*

Above:
Scorpions like this Vejovis *use their four pairs of legs both for walking and for digging a burrow in sand and gravel. This female is carrying her newly hatched young until they are able to fend for themselves.*

Opposite, top:
Millipedes like this desert Orthoporus *are built for strength and axial rigidity. Large numbers of short, powerful legs enable them to force their way into crevices without bending or distortion.*

Opposite, bottom:
*The sea slater (*Ligia*) is a marine relative of the familiar wood louse or sow bug, the most successful terrestrial crustacean. The curious way in which the legs of these creatures are arranged provides them with the locomotory benefits of long limbs but without the risks of having them extended far from the body.*

For species that utilize preexisting crevices, rather than excavate their own burrows, flexibility is a primary prerequisite. Symphylans, such as *Scutigerella,* are a good example of this type of morphology. Unlike diplopods, which have reduced their number of tergites through fusion, symphylans have added extra tergal plates at key locations along the body to provide improve mobility. For this reason, symphylans possess 14 tergal plates but only 11 actual segments and pairs of legs. A similar duplication of critical tergites is found in the extremely flexible New Zealand burrowing centipede *Craterostigmus.*

Unlike the vegetarian diplopods, most centipedes are active predators, and as such show adaptations for speed rather than power. As we shall discover when discussing the significance of different gaits, increased speed is achieved by the use of stepping patterns in which fewer legs are in contact with the ground at one time. This, in turn, tends to induce energy-wasteful side-to-side yawing. In powerfully built centipedes, such as *Scolopendra,* this lateral movement can be controlled by intersegmental musculature — albeit at the cost of increased weight. In *Lithobius,* a more lightly built but rapid runner, each segment and leg is associated with a single tergal plate, but in the midsection where yawing is most likely, segments 5, 7, 8 and 10 are enlarged to accommodate extra musculature to dampen out such lateral movement. The body of *Scutigera,* the most rapid and agile runner, bears only 7 tergites above its 14 pairs of long legs. This fusion of plates, combined with a complex musculature that unites each tergite with 5 sternal plates beneath, allows *Scutigera* to keep its body almost straight, even when traveling at high speed.

Arthropod Gaits

We are all familiar with trotting, cantering and galloping. These, together with walking, comprise the four gaits of a horse; these four patterns of stepping each evolved to enable the horse to achieve higher speeds. Similar variations in stepping pattern, geared to an increase in speed, are to be found in most terrestrial vertebrates, as well as in the onychophorans (velvet worms), myriapods and insects.

For quadrupeds the timing and sequence of leg movements is closely governed by considerations of stability. In creatures with six or more legs, this is less of a problem. However, those creatures with large numbers of legs must take special precautions against inadvertently tripping themselves up. As Mrs. Edmund Craster wrote:

> A centipede was happy quite,
> Until a frog in fun,
> Said "Pray, which leg comes after which?"
> This raised her mind to such a pitch,
> She lay distracted in a ditch
> Considering how to run!

For animals that depend upon fleetness of foot for their competitive edge, this can be a very real problem. Speed is gained by the use of long, fast-moving legs that employ an extended stride. In forms with many legs, this means substantial overlap of movement fields, which severely limits the available choice of gaits. This may explain the evolutionary trend toward a reduction in the number of functional legs in fast-moving groups.

This problem does not arise in *Peripatus,* however, because the legs are short and the fields of movement of adjacent legs do not overlap. Thus *Peripatus* enjoys the luxury of using whatever gaits it wishes. In fact, three gaits are so used: First, a low-gear, slow-start pattern that allows three legs to be lifted for every five on the ground. Second gear, an easy walking pace, involves elongation of the body, with 5 legs on the ground and 5 off it. Top

Viewed from beneath, this young Lithobius *displays the highly modified jawlike front pair of legs equipped with poison glands that are characteristic of centipedes. Large centipedes can inflict an extremely painful bite with these organs, which make them fearsome predators.*

RANGE OF LEG MOVEMENTS IN VARIOUS ARTHROPODS

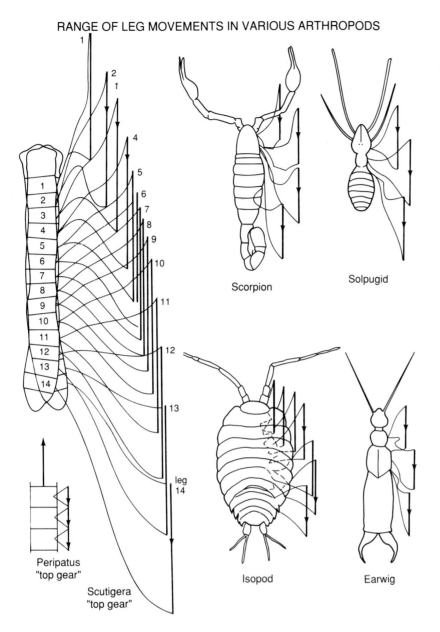

Scorpion

Solpugid

Peripatus
"top gear"

Scutigera
"top gear"

Isopod

Earwig

Next page, top:
The design of these walkingstick insects, photographed during mating at night, is influenced less by locomotory factors than by the demands of concealment. Nevertheless, they provide a good illustration of the basic insect body plan.

Fig. 19 Leg Movements in Arthropods. The heavy lines represent the movement of the tip of the leg in relation to the body during the propulsive backstroke. Scutigera, the most rapid of the centipedes, has substantial overlapping of movement fields, which allows it only one stepping pattern, or gait. Crustaceans and arachnids have not evolved the refinement of varying gaits as found in insects and myriapods.

gear, which does not necessarily produce a faster pace, but may be more economical, involves much greater extension of the body. Here five legs are lifted off the ground for every three in contact.

In *Scolopendra*, a moderately fast centipede with relatively short legs that do not overlap significantly, there is a choice of four different gaits, each with a greater number of legs interposed between the ones touching the ground.

For *Scutigera*, the fastest-moving centipede, which can reach speeds in the region of half a meter a second, the situation is very different. Special modifications to the coxal joint provide a greatly elongated promotor-remotor swing. Additional muscles attached to an inner projection of the coxa provide enhanced power to the leg movements. This in turn allows the use of greatly elongated legs, with considerable overlap (Fig. 19). The penalty for this enhanced performance is the elimination of all choice in the matter of gaits. *Scutigera* can use only a single gait, in which four out of the fourteen legs on each side are in contact with the ground at any given moment.

Next page, bottom:
Scutigera is the fastest of all the centipedes, and exhibits many adaptations associated with rapid running. These include enhanced range of leg movement and special strengthening to keep the body straight. The pattern of overlapping steps precludes the use of all but one gait.

JUMPING INVERTEBRATES

An ability to jump has evolved independently many times among invertebrates. In most cases, jumping is an escape reaction, rather than a normal mode of progression, as it is in such animals as jerboas and kangaroos. Although jumping can propel an animal rapidly out of immediate danger, most creatures who use this means of escape lack any way to stabilize themselves while airborne. As a result, landing is generally a haphazard, inelegant and time-consuming affair. While a series of repeated jumps can be distracting to pursuers, the direction of each one is generally random and uncontrolled. An exception to this rule is found in some of the primitive insects known as bristletails (Thysanura) in which pairs of legs are used to produce a repetitive, low-jumping gait. Likewise certain West African millipedes use an exaggerated coiling response to curl up the body and project it in the direction of movement as an escape reaction.

The height to which a jumping animal can leap is strictly limited by the laws of mechanics, and is proportional to the square of its velocity at takeoff. In order to jump to a height of one meter, antelope and grasshopper alike must each accelerate to 4.4 meters per second. Thus animals whose muscles perform with similar efficiency and occupy the same proportion of body weight should all be able to jump to the same height regardless of size. In practice this is not the case. Small animals suffer from two serious disadvantages. First, air resistance becomes an increasing problem with diminishing size, easily consuming more than half the energy applied to the jump in the case of a grasshopper. Second, the time in which acceleration can occur — the time during which the animal remains in contact with the ground — also decreases with size, rapidly leading to the need for relatively enormous forces to be generated in very short bursts of high intensity. Whereas a leopard need only produce forces of one and a half times its body weight to rise 2.5 meters, a flea jumping one-tenth as high must generate forces up to 200 times its body weight, which impose tremendous strain on both the skeletal system and on the internal organs.

Typically, a walking insect cannot exceed about 20 paces a second, which results in a top velocity of about 10 millimeters per second (mm/sc). For significant jumping to occur, a velocity of at least ten times this figure is necessary. Whereas a jumping locust takes about 23 milliseconds (ms) to reach its takeoff velocity of 3.4 meters per second (m/s), a flea has only about 0.8 ms to reach full speed — about 1 m/s. This is both faster than ordinary (nonfibrillar) muscle fibers can contract and beyond their power output in rapid contraction.

The solution to this apparent paradox, which affects all jumping insects, lies in energy storage and its sudden release through a catch or catapult mechanism. Fleas, click beetles, fly larvae, springtails and grasshoppers have all independently evolved a means of storing energy, generated through relatively slow muscle contractions, and releasing it suddenly by means of a catapult mechanism. We have already encountered energy storage by means of natural rubberlike proteins in the hinge joint of scallops, and we shall come across it again in the flight mechanism of insects.

In click beetles the catch mechanism consists of a peg and socket between the first two thoracic segments, and energy is stored by deformation of the body wall. The sudden uncoupling of the peg, as the beetle lies on its back, projects it up into the air. Several different kinds of fly larvae grasp the tip of the abdomen in their mandibles as muscle tension builds, possibly generating hydrostatic pressure against the elasticity of the body wall. Sudden release of the jaws causes the body to straighten and be thrown into the air.

Fleas use their hind legs for jumping. Powerful muscles within the thorax and coxae contract slowly before the jump, storing their energy in pads of the rubberlike protein resilin. These pads are derived from the resilin deposits that are normally used to help power the wing beat in flying insects. It is remarkable that the same structures should find this alternative locomotory use in insects that have lost the ability to fly. Slight movement about the roller-and-socket catch mechanism (analgous to that found in an electric light switch) at the end of the coxa allows the stored energy to be released virtually instantaneously. In grasshoppers the arrangement is somewhat different. The power for the jump comes from muscles within the femur, and storage is in the knee joint and tendons, but once again a catch mechanism ensures that the release of stored energy is very rapid.

It is the Collembola, or springtails, that possess the most extraordinary jumping mechanism. The jumping organ consists of a long projection from the tip of the abdomen, which is folded forward beneath the body and engaged in a hook on the third abdominal segment. Power to straighten the jumping organ is generated by muscular contractions deep within the body, stored in the elasticity of the body wall and transmitted by hydrostatic pressure.

Jumping spiders get their name from the habit of slowly stalking prey, and finally leaping on the victim once within range. The jump occurs when the hind legs are straightened. This arouses no curiosity until it is discovered that spiders possess no muscles at the knee joint that could cause such straightening. Like the jump of springtails, spiders also make use of hydrostatic pressure. This is generated within the cephalothorax, the toughened front half of the body, by contractions of muscles joining top and bottom. What remains unclear is how this very substantial pressure is prevented from reaching the soft, elastic abdomen. Although jumping performance in these spiders is modest by comparison with fleas, seldom reaching 25 times their body length, it is far more controlled. Power is generated appropriate to the distance from the target, be it prey or a remote twig, and in-flight stability maintained by a trailing silken dragline, which also prevents disas-

The peg-and-socket catapult mechanism of this click beetle is clearly visible as it lies on its back before takeoff. Devices of this sort are characteristic of jumping insects, which must generate forces hundreds of times greater than their body weight to gain the necessary acceleration.

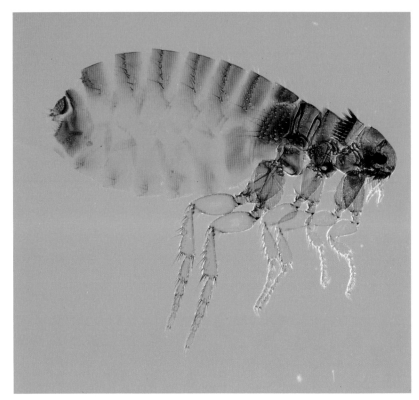

ter if the spider fails to reach its target.

The specialized hydraulic jumping mechanism of spiders is quite distinct from their normal walking behavior. As in insects, the power for walking in spiders and other chelicerates comes not from an active straightening of the leg, but from a promotor-remotor swing. What distinguishes the chelicerates, apart from their inability to utilize different gaits, is that the muscles that power the swing are inserted not on the coxa, which is immobile, but on the adjacent segment, the trochanter. For spiders, the trochanter has become a very important part of the body, despite its small size. Not only does it play this key role in movement of the legs, but it is also the site of special structures that allow the limbs to be autotomized. Autotomy is the ability to cast off a trapped or injured appendage without loss of blood. When the animal next molts, or sheds its skin, a fresh limb emerges from the stump. The complete new member develops within the trochanter and is inflated to full size by blood pressure before hardening.

LIMBS OF VERTEBRATES

The second great invasion of the land can be traced back some 375 million years to the osteolepid fish of the Devonian period. Although they had largely died out by the Carboniferous some 30 million years later, one osteolepid line has survived virtually unaltered to the present day, the famous coelacanth *Latimeria*. It was from ancestors of this sort that the early lungfish and amphibians evolved. We can see from the fossil record, and by comparison with living lungfish, how the pelvic and pectoral fins of osteolepids were first used for walking stiffly across the muddy bottoms of shallow lagoons. Once these fish began to emerge from the water briefly, in the manner of mudskippers, their fins were subjected to greatly increased forces. The resulting modifications to the skeleton and its associated muscles led directly to the amphibians and on to the reptiles. The posture of these early quadrupeds resembled that of present-day salamanders and lizards, with the long bones (humerus and femur) projecting horizontally from the body. Limb muscles were required to do little more than support the weight of the body, all propulsive effort being provided by powerful trunk muscles bending the spinal column. Indeed, such actions, inherited from the swimming movements of their fish ancestors, would have provided adequate propulsion even if the limbs had been rigid — a situation not unlike that found in polychaete worms.

The hind limbs of toads and frogs, the most familiar amphibians, have undergone considerable modification as an adaptation to swimming with both legs in synchrony and jumping. However, their slow walking movements in many respects resemble those of newts and salamanders, and have changed little from those of their early ancestors.

The early reptiles inherited the wide stance and powerful body movements of their amphibian forebears, but soon began to diversify and specialize. Flexibility was retained by lizardlike forms, some of which have

Jumping spiders do not have the necessary muscles to straighten their legs. Instead, the energy needed to jump comes from indirect muscle contractions that generate momentary bursts of high blood pressure within the legs, causing them to extend.

even lost their limbs and reverted to fishlike undulatory locomotion. This trend became most pronounced in the snakes. In total contrast, the turtles and tortoises encased their bodies in heavy armor and lost all mobility of the spine. For them powerful limbs became a necessity. Two ancient reptilian lines proved particularly successful, one leading eventually to the birds and the other to the mammals.

The limbs of mammals are oriented quite differently from those of reptiles. Instead of projecting horizontally to each side, the long bones of the leg, the humerus and femur, lie beneath the body and swing in a vertical plane parallel with the backbone. Although limb muscles provide the power for walking, in galloping and similar fast gaits, the action is heavily supplemented by trunk muscles flexing the spine up and down, which allows greatly increased strides. This flexure of the spine is particularly apparent in the marine mammals such as whales and dolphins whose swimming actions have been derived from the movement of terrestrial forebears.

Mammals occupy an enormous size range and many different life-styles. This diversity is reflected in limb structure and use. Small mammals tend to crouch, with legs bent, which keeps their center of gravity low and gives them the potential for rapid acceleration. In addition, many small mammals are active climbers. Their characteristic arrangement of limbs flexed beneath the body allows the hind legs to oppose the pull of the front legs, providing a secure hold on vertical and overhanging boughs. In this respect the limbs function like the tail feathers of climbing birds such as woodpeckers which are braced against the trunk.

The crouched stance of small mammals requires the leg muscles to be continually in tension, and would be very expensive metabolically for larger animals because of their proportionately greater weight. Thus large mammals tend not to climb and to stand with their legs straight. There are, of course, always exceptions. Some arboreal species hang upside down with their limbs stretched instead of compressed, which produces corresponding changes in both bones and muscles. Foremost among these are the sloths, which spend their entire lives inverted. Gibbons, which swing through the forest canopy by their arms—a process called brachiating— also show modifications appropriate to this change in stance and function.

Mudskippers (Periophthalmus) *use their front fins as primitive legs to pull themselves out of the water. Similar behavior by distant ancestors is believed to have given rise to early amphibians more than 400 million years ago.*

The lightly built limbs of this newt (Triturus) *project sideways from the body, resembling those of early fossil amphibians. Newts swim like fish by oscillations of body and tail. Their movements on land are clumsy and awkward.*

Stability

For an animal to retain equilibrium, it must have three legs on the ground at any moment in time, regardless of whether it is a quadruped or an insect. In addition, the center of gravity must lie within the triangle so formed. However, most moving animals do not remain totally in equilibrium, particularly when running. A quadrupd may periodically have all four feet off the ground at once. This does not matter if equilibrium is maintained overall from one stride to the next. Temporary lack of equilibrium will result in horizontal or vertical movements, which are corrected as the appropriate feet touch the ground, and without ill effect. But in slow-moving creatures such as tortoises, loss of equilibrium is potentially more serious. The variety of gaits available to an animal is directly related to the duty cycle of the legs, that is to say, the proportion of each stride that the foot remains in contact with the ground. On purely theoretical grounds it can be shown that tortoises, which have a duty cycle of about 0.8, have a very restricted choice of gaits if they are to remain in equilibrium. In fact, tortoises do not walk smoothly and economically, with constant equilibrium, even though this appears perfectly possible. Instead they pitch, roll and yaw appallingly, in what appear to be very poorly coordinated movements. The explanation lies in the fact that to maintain equilibrium at the gaits available, fast muscle movements are essential. However, tortoises, because they do not need to move fast, have evolved muscles of exceptionally high efficiency but which act very slowly. Through the use of mathematical models that mimic the action of tortoises walking, it can be shown that natural selection has adopted the best compromise possible.

Walking and Running

The speed at which an animal walks is governed by the length of its legs and the force of gravity. This is another manifestation of Froude numbers. It will be recalled that the Froude number was originally identified as a vital component of the drag imposed on a ship through the creation of a wake. Defined as *(velocity)*^{2} size × gravity, the Froude number also governs the

Alligators are ungainly on land and can walk or run only briefly. Considerable effort is needed to raise the body clear of the ground because the legs project sideways rather than downward.

Geckos are very agile lizards that have evolved special adhesive pads on their feet to provide greater traction. This enables them to climb without difficulty on smooth rocks and leaves.

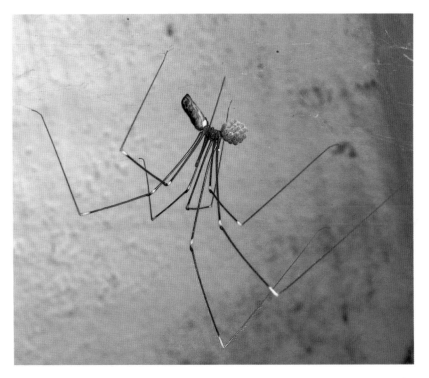

Unlike spiders that run on the ground, those that live in webs are built to hang upside down with their legs in tension rather than in compression. This daddy longlegs spider (Pholcus) is quite helpless if removed from its web. Note the bundle of developing eggs held in the fangs.

Opposite, top:
The body plan of chameleons, particularly the insertion of the legs, resembles that of alligators. However, being much smaller and lighter, chameleons have no difficulty climbing in trees and shrubs. Their movements are very slow and deliberate, in keeping with their cryptic lifestyle.

Opposite, bottom:
Alligator lizards (Gerrhonotus) belong to a family that also includes totally legless species. These long-bodied forms move in the same way as snakes, making relatively little use of their legs.

rate at which a limb can be swung back and forth. A human walker, for example, cannot bring his feet back to earth faster than the limit set by acceleration due to gravity. This means that at speeds above about 2.5 meters per second it is necessary for him to break into a run. This is a change of gait in which the duty cycle is shortened to less than 0.5 and the legs are bent at the knee, so shortening the time needed for the forward swing of the leg.

In the course of a stride the direction in which forces act on a foot changes. This becomes particularly apparent when one tries to walk on ice. As a foot is placed on the ground, force is applied at an angle running through it to the center of gravity at the hip. At the end of the stride, the force is applied backward. It is the reaction to this backward inclined force that drives the body forward. Not surprisingly, the effects of these loco-motory forces are often reflected in many fine details of a foot's anatomy. For example, a major advance in human evolution came about some 3 million years ago when hominid ancestors descended from the trees and began to walk over the East African plains on two legs. Among other benefits, this freed their hands for purposes other than walking and was undoubtedly instrumental in man's becoming a tool user. What can we tell about the bipedal history and abilities of these our forebears? The foot bones of Lucy, most famous of the australopithecines, suggest that her lineage had only recently moved out of the trees. Although the design of her knees, ankles and hip bones show that she was primarily bipedal, the curvature of her toe bones suggests that her feet were only recently used for walking rather than grasping. This interpretation is further supported by the configuration of the joints of the big toe, which were not strong enough to support the thrust inherent in a normal, full human stride, and indicates that Lucy and her kin shuffled around rather flat-footed.

Regardless of whether an animal is bipedal like Lucy, or walks on all fours, there are certain features of walking that are universal. As one foot completes its stride and pushes off, work is being done and energy expend-ed. As the next foot strikes the ground in front, energy is absorbed. This is negative work, but nonetheless necessitates the expenditure of energy. Both walking and running involve alternating cycles of positive and nega-tive work, and the same situation clearly occurs in kangaroos and other hopping animals as well. We have already noted that natural selection

Page 130:
The legs of climbing animals like squirrels are adapted for use in a wide range of postures. When the animal is clambering on vertical surfaces, as seen here, the hind legs are in tension. When it is moving horizontally along a branch the legs will be in compression.

Page 131:
The tree creeper (Certhia) hunts for insects on tree trunks. To maintain stability the powerful tail is used to brace the body in a vertical position, so helping the feet to maintain a firm grip.

strongly favors the conservation of energy, and consequently mammals have evolved very effective mechanisms to ensure that the negative work done as a foot strikes the ground is not wasted. This is done by temporarily using tendons and ligaments to store elastic strain energy. In this way energy is not squandered but used moments later as the limb recoils. This principle is found widely in all bipeds and quadrupeds, and can result in energy savings in excess of 30 percent. (See Fig. 20.)

Many factors influence the way in which a quadruped moves. Total size, as we have seen, directly governs the proportions of the limbs and also affects the extent to which a limb can swing. This is further influenced by how the mass of a limb is distributed and whether other ecological factors necessitate, for example, enlargement of the feet to facilitate prey capture or walking over soft surfaces. The form and proportions that an animal exhibits are the optimization of many conflicting requirements. For this reason we find that quadrupeds exhibit at least eight different running gaits — ambling, trotting, pacing, cantering, bounding and galloping, to name but some — even though the full significance of many remains obscure. Today there is a growing research effort aimed at trying to understand why a

Opposite:
Woodpeckers have even more strongly developed tails than tree creepers, and use them in just the same way. The woodpecker must be very firmly braced when digging out insects.

The slow muscle contractions of the box turtle (Terrapene) *are very energy-efficient. However, they make it impossible for the animal to maintain stability when walking. Consequently the movements of turtles and tortoises on land are very ungainly.*

Bipedalism has evolved independently in many groups of animals. It is the only form of locomotion available to flightless birds such as these ostriches, which can run at high speed for considerable distances.

particular animal moves in a specific way when performing certain functions. New findings are being reported at an increasing rate. What is becoming abundantly clear is that most parameters that used seemingly to depend on materials and mechanical engineering considerations are even more powerfully influenced by the demands of energy conservation.

LEG EXTENSION AND
SPINAL FLEXURE IN QUADRUPEDS

Thomson's gazelle (15 kg)

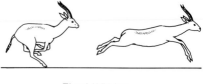

Fig. 20   Effect of Size on Running. A Thompson's gazelle, which weighs only about 15kg (33 lbs), shows much greater spinal flexure and extended stride than do larger antelope. The eland, weighing about 250kg (550 lbs), has a far more massive spine and is consequently less flexible. This gives the running eland a restricted, more stately gait. (After Bonner.)

Eland (250 kg)

Page 135:
Humans have evolved as strictly bipedal animals. This has affected the design of the whole body, including the orientation of the skull. These Samburu dancers use the elastic properties of the tendons in their legs and feet to jump high into the air without bending their knees.

Ground squirrels (Xerus) often adopt a two-legged stance to improve their ability to spot approaching predators.

Some animals are profoundly modified for particular life-styles. The mole (Talpa) spends its life in narrow burrows excavated with its powerful front legs, which are totally adapted for digging.

Above:
Zebras are close relatives of horses and are capable of covering long distances at speed. Predators such as lions must rely on stealth and ambush to bring down a zebra, which can easily outrun such attackers in the open.

Left:
Gray kangaroos (Macropus) are one of about 60 related marsupial species remarkable for their ability to progress fast and efficiently in enormous bounds. Using the legs in this way — simultaneously rather than alternately — is very uncommon except among frogs and jerboas. It is interesting to note that arboreal kangaroos move their legs alternately.

Next page:
Giraffes appear very ungainly, but can nevertheless move at surprising speed when necessary. A curious feature of giraffe locomotion is that both legs on the same side are moved together, as shown here.

Previous page:
Animals that climb sometimes use their tails as a fifth limb. The harvest mouse (Micromys) relies heavily on its prehensile tail as it clambers among stems of wheat.

Right:
Wildebeest have surprisingly slender legs for their size. They are typical grazing animals, designed for slow and steady progression except when attempting to escape predators.

Below:
Kudu are relatively large antelope and not well adapted for jumping, which causes a disproportionate stress on the leg bones as size increases. Small gazelles, however, can leap with impunity.

Left:
Baboons are equally at home on the ground and in trees. They normally move on all fours, but can run on two legs for short distances.

Below:
The steenbok has the typical proportions of a small, agile antelope.

Opposite:
Impala are fast-moving and agile antelope of medium size. Their ability to change direction suddenly in full flight often enables them to escape attacks by leopards and cheetahs.

Left:
Lions are large, powerful cats. Although strong, they lack the build to run fast. Indeed, they often steal prey from wild dogs and other smaller predators rather than kill for themselves. Lions will lie in wait and attack from cover with a short, swift burst of speed, but they will seldom try to outrun their victim.

Below:
Adult ground squirrels (Spermophilus) are immune to rattlesnake venom. To safeguard their young, the squirrels will divert attention to themselves, taunting approaching snakes by flicking dirt at them. The amazingly quick reactions of this squirrel enable it to leap backwards as the snake strikes, even while the dirt is still airborne.

5

FLIGHT

Page 144:
European blackbird (Turdus) *coming in to land.*

INTRODUCTION TO LIFE ON THE WING

Insects took to the air more than 100 million years before the first bird tentatively flapped its wings. This substantial head start in an environment free from competition has enabled the insects to diversify and refine their capacity for flight quite amazingly, and certainly well beyond the present limits of our understanding. It is prudent, therefore, that we depart from strict evolutionary chronology and begin our exploration of animal flight in slightly simpler surroundings by looking first at birds. There are two good reasons for this. First, the aerodynamics of bird flight are rather better understood than those of insects and may more readily be compared to the established principles of aircraft flight. Second, the diminutive size of many flying insects makes them particularly susceptible to the influence of small Reynolds numbers, which further confuses an already complicated story.

The underlying aerodynamic principles that make flight possible were discussed at some length in Chapter 2, and it is now time to look more closely at how these principles have been invoked and exploited by the forces of natural selection. It is often difficult not to regard the exquisite adaptations of form and function that we observe in living organisms as being anything short of perfection. However, what may appear to be perfection is in reality nothing more than marvelous compromise. Natural selection can only mold existing genetic clay, modifying structures and systems previously honed to other uses. This applies with particular relevance to the evolution of flight. All flying animals have evolved from terrestrial forms, which in turn underwent profound modification in the descent from their aquatic ancestors. Imagine the problems an aircraft designer would face if asked to modify a wheeled submarine for flight! When viewed in this light, natural selection has worked some truly remarkable alchemy.

THE ORIGIN OF FLIGHT IN VERTEBRATES

The first vertebrates to venture into the air were reptiles. Called pterosaurs or pterodactyls, they first appeared about 190 million years ago, and persisted until the end of the Cretaceous period. This was 63 million years ago — about the time when the Rocky Mountains first began to rise. These much maligned creatures include the largest fliers known — giants with a wingspan of over 10 meters and weighing some 90 kilograms — but many were much smaller, resembling seagulls and sparrows in size. There has been a lot of speculation about the flight of the large pterosaurs, such as *Pteranodon* and *Rhamphorhynchus,* culminating in the construction of a half-size flying replica of the giant Texan *Quetzalcoatlus.* From this it appears that pterosaurs were nature's precursor to the hang glider. Lacking the massive muscles necessary for takeoff, these ungainly creatures were obliged to launch themselves from nearby cliffs, but once airborne could maneuver well and ride the thermal updrafts. Like so many early human aviators, the pterosaurs were not terribly successful, and have left no modern descendants. That is not to say the experiment failed. After all, pterosaurs lasted for over 100 million years, which is 20 times longer than man has been on earth!

The end of the pterosaurs coincided with the end of the age of dinosaurs 63 million years ago. Whether or not this period of mass extinction was precipitated by some cosmic catastrophe such as collision with a giant

meteorite, the pterosaurs were already doomed. They were facing increasing competition from an unrelated group of reptiles that were evolving into birds. The earliest birdlike creature of which we have good fossil evidence is *Archaeopteryx*. Dating from about 170 million years ago, several *Archaeopteryx* fossils have been recovered from fine-grained Jurassic limestone in Bavaria, and these have provided a wealth of anatomical information. Who their ancestors were is a matter of paleontological controversy, but it appears likely that *Archaeopteryx* arose from small, agile warm-blooded arboreal dinosaurs. In many features such as teeth and tail, these fossils are clearly reptilian, but in others, such as claws and feathers, they already possess many characteristics that are typically avian. Of special interest is the structure of the breastbone, or sternum. As those who have carved turkey or chicken know, the massive flight muscles of birds, which comprise the breast meat, are divided by a powerful keel protruding from the sternum. This provides the skeletal anchor point for these muscles, and its size gives an indication of flying power. *Archaeopteryx,* despite its feathers and other birdlike characteristics, does not have a keeled sternum. Thus it could only have been capable of gliding down from trees or small cliffs, possibly to escape approaching predators. It is easy to envisage the gradual improvements that were imposed upon this basic body plan, leading in time to the emergence of fully functional, typical birds. By the Eocene epoch, 54 million years ago, the fossil record contains an abundance of bird remains.

Some birds are specialized in directions other than flight. This African jacana (Actophilornis) spends much of its time searching for food among lily pads. The greatly enlarged feet help spread the load and prevent the bird from sinking.

This body feather from a tawny owl (Strix) provides insulation as well as contributing to the smooth surface contour that allows for silent flight.

FEATHERS

Of all the remarkable flight adaptations possessed by birds, none perhaps is more extraordinary than that of feathers. These are derived from ancestral reptilian scales, and have diversified in form to serve a variety of different roles. The skin of reptiles is characteristically dry, and contains few glands, unlike that of amphibians or mammals. The outer surface is covered in horny scales, which are periodically shed. Like mammalian hair, the scales of reptiles consist of keratin, a rigid protein of characteristic structure. Its great strength and rigidity come from the extensive cross-bonding between the polymer chains of which it is composed.

Unlike other structural elements such as bone or insect cuticle, which are secreted by cells, keratin is laid down only within cells. Hence reptilian scales, bird feathers and mammalian hair are all composed of dead cells, tightly packed with keratin. Reptilelike scales are still found on the legs of birds, but over the rest of the body the scales have been replaced by regularly distributed tracts of feathers.

The main support of a typical wing feather is its central shaft, or quill, which is hollow for much of its length. From the quill project numerous barbs, each bearing, in turn, large numbers of little filaments, or barbules. Minute hooks on the barbules unite the barbs to form a continuous web, structurally strong, but very light and economical of materials. As every child who has ever played with feathers knows, the connections can be easily broken so that the feathers separate, but preening equally easily reunites them.

A large bird may possess over 20,000 feathers, those from each part of the body characteristically unique in form. Despite their individual lightness, their combined weight often exceeds the weight of the entire skeleton. This is not so surprising when one considers that feathers not only provide the principal aerodynamic surfaces for generating lift, but also envelop the body in a regular, streamlined covering. One has only to look at the naked, inelegant ducks decorating a Chinatown shop front to appreciate their importance in this role.

Beneath their sleek, aerodynamic outer feathers, birds are covered by a dense layer of soft down, whose purpose is to provide thermal insulation to keep the body temperature at about 46 degrees C. Indeed, it is thought that this was the original role of feathers in their evolution from scales. In addition to the down, flight and contour feathers, birds can also possess several other specialized types. These include hairlike sensory filoplumes, eye lashes and feathers used for ornament and display like the peacock's tail. As we shall see, the shape and arrangement of feathers figure prominently in discussions of specialized flight adaptations.

The striking patterns of this peacock feather are interference colors and are caused by details of fine surface structure rather than pigmentation. The role of these feathers is now solely for male ornamentation in courtship.

The Silent Flight of Owls

Many birds generate considerable noise when they fly, but owls are a notable exception. Indeed, close examination of an owl's feathers reveals a host of specializations that appear to be specifically designed for silent flight. This is usually assumed to improve their performance as hunters by allowing them to approach their prey unheard. However, it is much more probable that reducing their own noise allows them to make maximum use of their very acute hearing. This would enhance their excellent night vision and improve performance in hunting small mammals.

First, the wing feathers bear a fine down on their upper surface, so that they move silently over one another. They may also provide a compliant surface to occlude incipient turbulence, which would also minimize noise. Likewise, turbulence immediately behind the wing is suppressed by a fringe of small extensions from the hindmost feathers.

The leading edge of the owl's wing is remarkable for a row of curious projecting barbs on the first and second primary feathers that resemble little hooks. They, too, are assumed to contribute to the owl's silent flight, but as yet there is no convincing explanation of their function. The barbs are about one-tenth of a millimeter apart and clothed with a fringe of fine

Seen in close-up, this peacock feather shows the way in which a feather is made up of very fine interlocking barbules that fill the spaces between the barbs, the parallel elements that project from the central shaft.

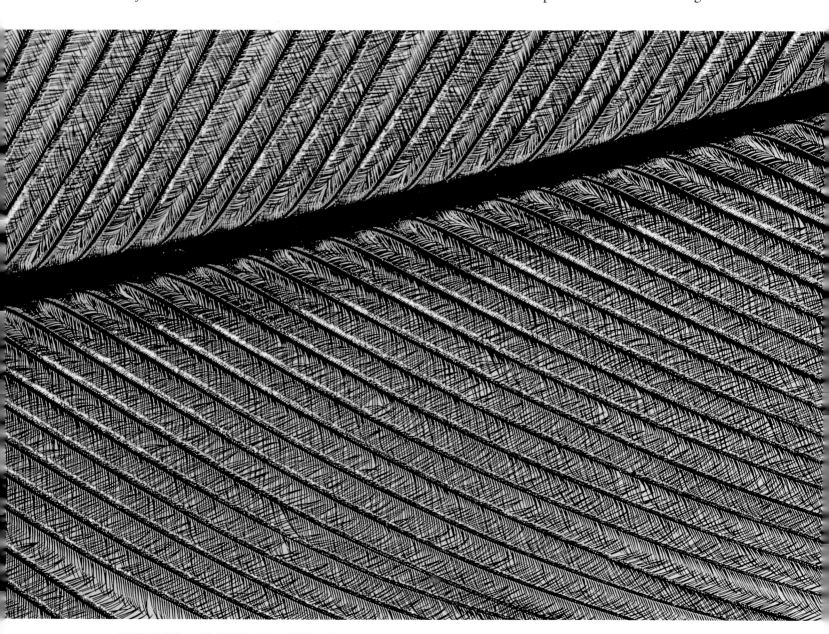

This barn owl (Tyto) is gliding between wing beats. Notice the leading edge slot formed by the alula, which is clearly lifted away from the wing.

barbules on one side. This brings them into a Reynolds number range of less than 100, in which flow would be very viscous. It is possible that the barbs in some way influence the boundary layer and improve lift at high angles of attack or under high wing loadings. They do not appear to function like the spoilers or vortex generators on the wings of jetliners, however, because their orientation to the airflow is inappropriate for this.

FURTHER FLIGHT ADAPTATIONS

Structural modifications associated with flight are not confined to the outer surface, however. The whole body of a bird reflects the extreme demands of flying. Thus the skeleton, originally evolved for walking, has undergone striking changes. The bones of the pelvic and pectoral girdles are notably modified, the latter by the enlargement of the sternum to support the massive flight muscles. However, the enlargement and strengthening of certain skeletal elements has been accompanied by a rigorous exclusion of all unnecessary weight. Consequently, bones have become hollow and porous, retaining material only in those regions where strength is essential. Likewise many individual bones have fused together to form lighter, stronger units. Even the skull has not escaped such adaptations. Heavy reptilian teeth have been replaced by a light, horny beak, and the enhancement of visual acuity at the expense of smell is also reflected in the relative size of various regions of the skull.

The powerful form of the chest cavity in flying birds reflects several important flight adaptations. Not least among these is the evolution of a specialized respiratory system. In mammals, the flow of air to the lungs is intermittent as the animal breathes in and out. Birds have evolved a more efficient system in which airflow through the lungs is continuous and about 90 percent of the available oxygen extracted. This is made possible because in the lungs of birds the air passages do not end in blind alveoli, but divide into numerous pores, or parabronchi.

Associated with the lungs are some large, smooth-walled sacs, which were first described in 1653 by William Harvey (1578–1657), who, it will be recalled, also discovered the circulation of the blood. Three hundred years were to elapse before the function of these sacs was understood. Air is drawn into the posterior sacs through the trachea, and passed forward in a steady stream through the lungs. From the lungs, air passes to the anterior sacs and from them to the outside, again via the windpipe. Blood and air flow in opposite directions, with the oxygenated blood about to leave the system encountering the freshest air. This improvement in lung efficiency is particularly significant for those species that migrate at high altitude. Humans and other mammals exhibit significant oxygen deficiency above 10,000 feet. Birds, however, can function well at 20,000 feet, where there is only half the amount of oxygen present at sea level.

This section through the femur of an ostrich (Struthio) shows how even in flightless birds all unnecessary weight has been eliminated. This bone is full of air-filled spaces and only ossified where it is needed for mechanical strength.

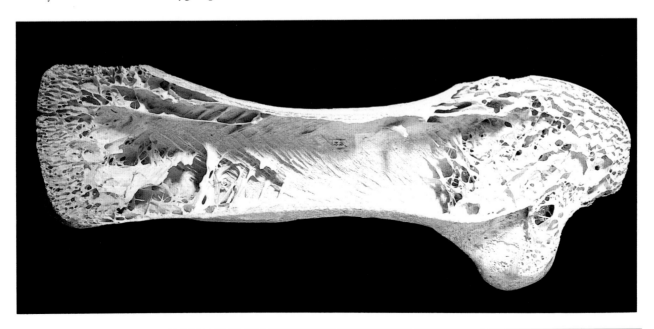

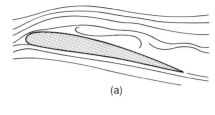

(a)

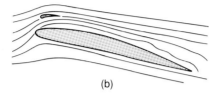

(b)

Fig. 21 The Effects of the Alula on Airflow. A wing that is heavily loaded risks stalling if the airflow over it starts to separate and become turbulent (a). The alulae of birds (b) funnel extra air to the lifting surface, effectively limiting turbulence to the hindmost part of the wing, where it has little effect on lift.

The fish eagle (Haliaetus) plucks fish from beneath the surface as it swoops low over the water.

WINGS

As anyone who has sat in a window seat on board a modern jetliner will have noticed, man-made wings are quite complicated, being embellished with an array of flaps, slots and assorted protuberances. We can find close functional parallels between such structures incorporated by engineers into aircraft wings and those present in bird wings, evolved through natural selection.

In order to maintain lift at low speeds, modern aircraft have slots along the leading edge that can be opened for landing and takeoff. These slots channel more air over the upper surface of the wing to delay the onset of turbulence and confine it along the trailing edge. This allows greater angles of attack, and so greater lift, without the risk of stalling. In birds, the bones that equate to the human thumb are also used to form a leading-edge slot (Fig. 21). Called the alula, or bastard wing, it is positioned along the front of the wing and opens during landing and takeoff in just the same way. In two groups of birds, however, the alula is absent. Among the loons, which are primitive, inefficient fliers, the need for such sophistication has not yet arisen. In contrast, the flight of hummingbirds has become so specialized that the alula has ceased to be of benefit. Having served its purpose and become redundant, the alula of hummingbirds has been eliminated in the course of evolution.

Viewed from directly beneath, the arrangement of feathers at the wing tips of many birds appears somewhat ragged. When compared with the smooth contours of an aircraft wing, it might be thought that such an arrangement would prove disadvantageous. However, evolution is once again providing an elegant solution to a particular problem. It will be

recalled from Chapter 2 that the difference in pressure between the upper and lower wing surfaces causes air to seep around the tips, giving rise to the wing-tip vortices that are responsible for induced drag. In addition, lift is no longer generated from the region where the vortices form. This is particularly disadvantageous on heavily loaded, stubby wings like those of owls. In such birds we find the feathers precisely sculpted to form wing-tip slots that reduce the size of the trailing vortices and hence the region of diminished lift they cause. But this is not all. The wing tips are seemingly subjected to a number of other special problems for which the pressures of selection have evolved solutions. In tight turns, for example, airflow and lift vary significantly in different parts of the span, and there is a very real danger of stalling at the tips. In man-made aircraft subjected to this problem, such as agile agricultural crop-sprayers, accessory winglets are added to the wing tips. In birds such as vultures, who spend much time circling in thermals, a comparable solution is found. The feathers of the wing tips are structured to permit a degree of bending, in addition to forming slots. At the reduced Reynolds numbers at which they are functioning (because of low velocity) they are able to generate an exceptional amount of lift at the tips with minimum drag. The flexible, porous nature of the wing surface, resulting from its covering of feathers, doubtless leads to numerous ways of controlling flow, although these are not yet understood. For example, it may be that the rippling of feathers observed over the wings of soaring birds is not simply the result of unsteady flow, but a specific response to control turbulence, akin to the vortex generators on some jetliner wings. These are the small, angled vanes projecting from the upper wing surface just behind the leading edge. By promoting the formation of a series of counter-rotating vortices across the wing, thinner, more stable boundary layer conditions are created between the vortices. The result is a dramatic delay in separation, with a corresponding gain in lift.

Tropic birds (Phaethon) *are specialized marine soarers, and have long, narrow wings similar to those of the albatross. However, they belong to different orders, the tropic birds being more closely related to the pelicans.*

Hang glider pilots fly suspended beneath their wings. Because they are much heavier than their aircraft, only slight movements of the control bar are needed to change attitude or to turn. The aerodynamic design of a hang glider makes it a remarkably stable machine to fly.

Who knows what other sophisticated flow control devices remains to be discovered? The bird's wing, being a flexible, mobile structure whose shape and area are continuously variable, and whose surface is also controllable in both form and texture, is extremely difficult to study under laboratory conditions. Early attempts to study the aerodynamics of bird wings in wind tunnels, using dead birds, were a singular failure. All they showed was that dead birds cannot fly!

THE ART OF SOARING

A bird gliding at constant speed is in equilibrium. That is to say, the forces acting on it—lift and weight, thrust and drag—cancel each other out. Weight acts vertically downward through the center of gravity. The aerodynamic forces of lift and drag act through a comparable point in the wing called the center of pressure, which in equilibrium is directly aligned with the center of gravity. By slight movements of the wing, the center of pressure can be shifted, so altering the tilt of the body and the angle of glide. Hang glider pilots achieve the same effect by shifting their weight relative to the wing. By this means the bird can adjust its attitude for whatever function is needed. In migration, for example, it may be advantageous to try to cover the greatest possible distance, while in hunting it may prove best to fly at the minimum sinking speed in order to remain aloft for the longest possible time. For landing it is obviously best to approach at the lowest speed without stalling. For each of these strategies there is an optimum gliding angle.

Powered flight is an expensive method of locomotion and is inescapably burdened with high energy costs. If the flight of eagles or vultures appears to be almost effortless, it is because these birds have discovered how to exploit natural air movements. This is a special class of gliding known as soaring. The most important of these natural air movements are thermals, rising air currents generated through the uneven heating of the ground beneath. Although the detailed structure of thermals is still unclear, in essence they consist of a series of swirling, doughnut-shaped vortices that rise up thousands of feet into the surrounding air. These are quite different in structure and energy to the swirling maelstrom of a tornado or dust devil.

By circling within the vortices of a thermal, a bird's sinking speed is less than the upward movement of the air, which over the great plains of East Africa has been measured at about four meters per second. After gaining altitude within a thermal, the bird then leaves it and glides earthward, covering ground until it can locate another thermal and once more begin the climb without effort. Using this method, which is exactly the same as that used by glider pilots, birds can cover enormous distances. A Ruppel's griffon vulture was followed by motorized glider over the Serengeti Plain for 75 kilometers, which it covered in just 96 minutes, after climbing in only 5 thermals. These birds often travel twice as far when they fly regularly from their nesting sites to distant feeding grounds — all without the expenditure of significant amounts of energy. Whereas glider pilots must rely on the presence of soaring birds and telltale clouds to locate thermals, it seems

An American egret (Casmerodius) *flying through the Florida Everglades. Although egrets hunt by standing immobile for long periods, they are good fliers and migrate considerable distances.*

likely that the birds themselves have a natural detector system, using low frequency sounds as a means of locating regions of turbulence.

Among other birds that rely on thermal soaring for long-distance travel are the storks that annually migrate between Europe and Africa. However, thermals do not form over water because it heats more evenly, and so the storks are obliged to follow a roundabout route, tracking the appropriate land contours around the eastern end of the Mediterranean Sea.

As with so many other aspects of locomotion, thermal soaring involves compromise solutions between conflicting demands. The best soarers should, theoretically, be large birds with long, narrow and lightly loaded wings—a condition that is generally found in sailplanes, and is typical of the albatross. However, the forces acting on very long wings do so at a considerable distance from the bird's body. This necessitates larger, stronger muscles and appropriately strengthened skeletal elements. These in turn impose an increased weight penalty. Moreover, the problem is further exacerbated by the flight pattern that thermal soaring imposes. The tight circles often necessary to keep within the thermal generate centripetal forces that join gravity in opposing lift. Thus the wings of terrestrial birds that employ thermal soaring are shorter and stronger in proportion than those of the albatross. However, the albatross is able to use a specialized soaring technique available only to sea birds.

Opposite:
Rainbow lory (Trichoglossus).

Gradient Soaring

Perhaps the most elegant gliding technique of all is that used by the albatross *Diomedea,* which spends much of its life soaring over the surface of the open ocean. Known as gradient soaring, it relies upon the fact that the wind blows more slowly close to the water surface than at 40 feet above it. In fact, this wind-speed gradient is a boundary layer effect, similar in all but scale to that discussed in Chapter 2.

The albatross flies in a characteristic, repetitive pattern. From a height of 25 or 30 feet above the ocean surface, it turns downwind and glides toward

The flight of the albatross (Diomedia) *is the epitome of soaring perfection. By playing the subtle changes in air currents across the ocean surface, the albatross can remain airborne for days with very little expenditure of energy.*

Fulmars (Fulmarus) are typical inhabitants of sea cliffs. They are soaring birds that are adept in exploiting the updrafts that abound in their domain.

the water, gaining both speed and kinetic energy in the process. As it nears the surface, it turns sharply into the wind and begins to climb. As it does so, its newly acquired kinetic energy is converted back into potential energy. Seen by an observer in a boat, the bird's velocity appears to decrease as it climbs. However, its airspeed — the only thing that really matters — remains constant as it progresses up the wind gradient. Repeated soaring in the wind gradient, with an occasional boost from updrafts generated over the crest of the ocean swell, enables the albatross to remain airborne for hours and even days with hardly a flap of its enormous wings. Indeed, it has been found that through gradient soaring, albatrosses are able to circumnavigate the globe, traveling a distance of over 30,000 kilometers in about 80 days — an achievement that would surely have gratified Jules Verne!

Slope Soaring

Another important energy-saving technique employed by many birds is slope soaring. This involves making use of local updrafts created by hills, cliffs and other surface irregularities. In fact, we have already encountered a special form of slope soaring, namely the use made by albatross, petrels and other sea birds of the shifting currents of air over the crests of waves. Slope soaring is widely practiced by raptors such as kestrels and falcons as they hawk across hillsides in search of prey. To be successful as a soarer, it is advantageous to be able to glide at relatively high speeds, but at a low angle or sink rate. Only large birds with high-aspect-ratio wings can glide at low angles, while high speeds at low glide angles dictate high wing loadings, which are a characteristic of short, broad wings. In order to maximize these conflicting requirements, birds have the ability to alter quite markedly the span and surface area of their wings under differing flight conditions. But, as usual, for most birds the best solution is one of considerable compromise. Sea birds such as gulls and fulmars make extensive use of slope soaring as they patrol the cliffs upon which they nest. However, such species cannot aspire to the theoretical design optimal for soaring because of their nesting behavior. In order to land safely on small ledges, it is essential that their wings function also at low stalling speeds. Thus cliff dwellers tend to be intermediate in both size and aspect ratio.

Insects fly at much lower speeds than birds, and in general do not make much use of gliding flight. However, it is significant that migratory species such as monarch and painted lady butterflies and the migratory locust are able to glide extensively. Butterflies with this ability are often to be seen slope soaring close to buildings in windy conditions.

POWERED FLIGHT

In considering the powered flight of birds, it is convenient to make a distinction between fast and slow flapping, because they involve rather different aerodynamic principles. Hovering and slow flapping, when forward motion is by definition less than the speed of the wing tips, will be discussed in a later section, along with the peculiar flight of hummingbirds.

Although flying birds range in size from 6 grams to over 9 kilograms and cruise at speeds of between 12 and 40 meters per second, the wing strokes that propel them are similar in all species except hummingbirds. Detailed analysis of the forces acting on a wing is extremely difficult because the geometry is continually changing and the flexible, porous nature of the feathers has an unknown influence on airflow. Nevertheless, it is clear that the aerodynamic forces acting on a bird's wing are comparable to those encountered on the wing of an aircraft. This is not so with insect wings. The tip of a typical bird's wing moves in a flattened figure eight, twisting about its long axis to present a varying angle of attack. The overall effect is that the upstroke is one of recovery, adding little to the bird's motion, and that the downstroke generates the necessary aerodynamic forces for flight.

A major difference separating the flight of birds from that of aircraft is stability. As the Wright brothers realized, a man-made flying machine has to be inherently stable, with any departure from straight, level flight being self-correcting. This is because man is not endowed with the necessary sensory equipment to detect and analyze incipient instability, or the coordination to initiate an appropriate response to correct it. In birds, and even more strikingly in insects, the whole flight system has evolved in unison, complete with mechanisms for correcting instability. Why is this important? Stability in flight must be paid for through a corresponding lack of maneuverability, and in the competitive climate of the real world, natural selection places a high premium on agility. Whether to catch prey or evade being eaten, low stability and high maneuverability are enormously advantageous.

In aircraft, the relatively small control surfaces of the tail plane are able automatically to counteract incipient deviations from the flight path generated through variation in airflow over the wings. This is so because the relatively small corrective forces generated by the tail plane are magnified because of the length of the fuselage. Insects generally lack control surfaces analogous to a tail plane, but this is not so in birds. Although not inherently a stable system, the tail feathers of a bird provide valuable control surfaces, serving for both steering and modifying lift. When a bird is coming in to land, its legs are swung forward, altering the bird's center of gravity, its attitude and its pattern of drag. The tail feathers are also deflected downward, changing the camber of the aerodynamic surfaces to provide braking and reduce the stalling speed.

Because flapping flight demands high energy consumption, there is strong selective advantage in favor of structures or behavior patterns that contribute to economy. For example, small birds often fly with a characteristic bounding flight, closing their wings close to the body for brief periods between flapping. It can be shown that for small birds that maintain relatively high speeds, this method of intermittent flying saves energy by periodically reducing drag.

Similarly, the V-formation adopted by skeins of migrating waterfowl, and so beloved by wildlife painters, is no accident. The leading member of the flight generates a turbulent wake, particularly through induced drag. This, it will be recalled, is formed by wing tip vortices as high-pressure air beneath the wing creeps round to the low-pressure zone above. The birds behind the leader position themselves to exploit the lift produced by these vortices. Similar behavior by succeeding members of the flock spaces them out in a regular pattern. The savings in energy produced by this behavior depends on the accuracy with which birds are able to maintain station on one another. Correct positioning within close tolerances is critical to gain maximum benefit, which can be 36 percent over solo flying. Of course, such benefits do not accrue to the lead position, which requires the consumption of substantially more energy than other places in the skein. What motivates an individual to assume this position is not clear. Even though the leadership of a migrating skein changes quite frequently, only a small number of individuals in the flight ever move into the lead. This behavior is unlikely to be altruism, but presumably confers some compensatory benefit such as improved status and corresponding mating success once the migration is over.

HOVERING FLIGHT

Slow flight and hovering flight present special aerodynamic problems, particularly to large and medium-size birds. It can be shown that larger species require proportionately more energy to hover than do smaller ones. The level of oxygen consumption becomes so great, in fact, that only the smallest species—namely the hummingbirds—are able to hover continuously. Pigeons, which are among the best medium-size hoverers, are able to remain in one spot for only a few seconds.

Observation reveals that birds use rather different hovering techniques depending on their size. Among the smaller passerines—that is, the common perching birds—hovering flight resembles normal forward flight in that lift is generated only on the downstroke. On the return upstroke, the wing is rotated so as to present minimal wind resistance. For pigeons and birds of similar size, hovering is so costly and difficult that a quite different type of wing movement has evolved. On the upstroke, the wing is rotated so as to produce a high angle of attack to the air through which it is traveling. Under normal circumstances this angle of attack would be well above the stalling angle, but modifications to the design of the primary feathers prevent stalling. Unlike most feathers, the shaft runs not down the center, but about one-third of the way back from the leading edge. In hovering, each primary feather rotates, opening the wing like a venetian blind and generates lift individually. By this means the wing is able to contribute power on both the up- and the downstroke.

The most specialized hovering flight, as already indicated, is found in the hummingbirds, a group confined to the New World. Birds filling the same ecological niches in other parts of the world, such as the sunbirds in

Cuckoo (Cuculus) *in flight. The wing is near the bottom of the downstroke and the alula is clearly visible.*

Africa, are unable to hover like hummingbirds, even though they are of similar size. The hummingbirds' success depends upon their ability to generate lift efficiently on both up- and downstrokes of the wing. In fact, the terms up and down in this context, although anatomically correct, are misleading. In hovering, hummingbirds hold their bodies vertical, so that the wings beat horizontally. In the course of each cycle, the wings are strongly rotated through almost 180 degrees. This is only possible because of their exceptionally mobile shoulder joint, all part of a very advanced evolutionary package. The hovering flight of hummingbirds has often been likened to that of a helicopter with a reciprocating rotor. As explained in Chapter 2, this analogy gives a false picture. For all birds, and hovering insects as well, the lift forces are the result of a momentum generated by the formation of vortex rings.

Takeoff and Landing

For birds that are able to hover, takeoff presents no special difficulties. Once airborne, they are soon able to build up sufficient forward speed for normal flight. For those unable to hover in one place, alternative strategies must be found. The first essential is to generate sufficient lift at low forward speeds without stalling or losing directional control. Aircraft can build up speed gradually as they accelerate along the runway until the lift generated by airflow over the wings exceeds the force of gravity. Large water birds

Swans (Cygnus) are among the heaviest birds that can fly. Takeoff and landing are particularly difficult in still weather as considerable wind speed is necessary for such large animals to become airborne.

such as geese and swans employ a similar technique, bounding across the surface until they are moving fast enough. For those that live in trees or on cliffs, it is only necessary to step out into space. Gravity ensures rapid acceleration and the bird is soon airborne in a graceful downward swoop. Frequently the wind is already blowing sufficiently strongly to allow the bird simply to spread its wings to become airborne.

Two thousand years ago, the Latin poet Virgil commented on the distinctive clapping sound made by pigeons on take-off. This is caused by the wings coming together violently above the bird's head. It has recently been suggested that this noise has considerable aerodynamic significance. As we shall soon discover, flying insects make use of some curious aerodynamic principles that depend on unsteady flow. One such technique is called "clap and fling." First described in tiny parasitic wasps functioning at very low Reynolds numbers, the same mechanism is also used by much larger butterflies. It is possible that pigeons too are using a similar technique at takeoff.

Like takeoff, landing too presents special problems. Many water birds come in to land without reducing speed, cushioning the impact with their feet and skidding to a halt. This can lead to a certain loss of dignity if the water is frozen! For most birds, landing calls for delicate timing and coordination. Like hang glider pilots, birds must swoop down to their chosen landing site without losing so much speed that they stall. At the appropriate moment, the wings are rotated beyond the stalling angle — "flaring out," in hang glider terminology. The sudden loss of lift and greatly increased drag allows the bird to drop gently onto its perch. As many novice hang glider pilots painfully discover, it is not always easy to judge speed and altitude accurately in these circumstances.

FLYING MAMMALS

The world's most diverse and sophisticated fliers are the insects, which represent the culmination of more than 350 million years of selection and evolutionary refinement. In contrast, the bats have only been airborne for about 50 million years. Even so, their flying abilities are impressive, as demonstrated by their skill at capturing insects on the wing. Although bat species are outnumbered by birds nine to one, this reflects a remarkable level of success in competition against a group who had already held command of the skies for 120 million years and were able to exploit a wide diversity of ecological niches. Their success is even more apparent when viewed against the evolutionary radiation of the mammals as a whole. Bats represent almost a quarter of all mammal species.

The aerodynamics of bat flight is not as well studied as that of either birds or insects. Consequently, the elegant adaptations that we have seen for controlling airflow over the wings of birds have not yet been identified in bats. Indeed, it may well be that bats have not yet been pressured into refining the aerodynamic aspects of their flying in the same way that birds have. Through the development of a highly sophisticated echolocation or sonar system that allows bats to fly at night, they have largely managed to avoid direct competition with birds, most of whom are strictly diurnal. Instead, selection in bats has tended toward improvements in their sonar system and increased maneuverability.

The largest bats, with wingspans of up to 2 meters and weighing 1.5 kilograms, are the flying foxes. Unlike other bats, which belong to the

*This male Anna's hummingbird (*Calypte*) is hovering. To do this its wings are oscillating horizontally rather than vertically. The hummingbird shoulder joint allows the wings to be rotated so strongly that lift is generated in both directions.*

The aerodynamic features of the long-eared bat (Plecotus) *are somewhat overshadowed by its enormous ears, which presumably are specialized in connection with its sonar system.*

Microchiroptera, flying foxes have large, well-developed eyes and no sonar navigation system. Their wings are of typical bat construction, with a fine membrane of skin, enclosing bundles of muscle and elastic fiber, stretched between the greatly elongated bones of the hand and across to the body. However, because flying foxes have no tail, the flight membrane is not carried back between the hind limbs as it is in other bats. There is considerable variation in the shape of the tail in Microchiroptera, and this appears to be correlated in some way with flying behavior. However, beyond being used to capture or temporarily store flying insects, the exact role of the tail membranes in flight remains unclear.

Bat wings, which generally beat at 15 to 20 Hertz in normal flight, generate lift only on the downbeat, and produce a wake of small vortices. The upstroke, however, can take several forms and fulfill a variety of roles, depending on flight pattern and speed. In slow flight the tips of the wings are flicked backward at the end of the recovery stroke to provide intermittent forward thrust. This forward thrust generates alternate vortices in the wake, opposite in sense to those produced by the downbeat. At higher speeds, the upbeat is entirely unloaded, and forward thrust has to be generated along with lift on the downbeat. A third type of wing movement is employed for cruising flight with minimal power consumption. This movement involves a marked reduction in wingspan on the upstroke. Curiously, this reduction actually results in the creation of negative thrust for part of the cycle, which is more than compensated for by the positive forward motion generated during the powerful downstroke, in which the wings are fully extended.

Although bats are unable to alter significantly the span of their wings and the loading on the powered downstroke — an important gliding adaptation found in birds — they possess considerable control over camber and twisting, which is important for agility. Variations in aspect ratio can be correlated with differing life-styles. Species with long, narrow high-aspect-ratio wings, such as free-tailed bats, are fast, high fliers in habitats lacking obstructions. Most migratory bats, some of whom regularly travel up to 800 kilometers, covering 50 kilometers or more a day, have high-aspect-

The echolocation system of bats is remarkable, particularly the ability it gives them to capture fast-moving insects on the wing. Noctuid moths, like the one here, have evolved the ability to tune in on the bat's sonar frequency and take avoiding action.

ratio wings. In contrast, short, broad low-aspect-ratio wings, like those of horseshoe bats, are typical of species that need to fly with great precision through dense foliage.

Anatomically, bats are not markedly different from other mammals except for the lightness of their skeletons and the structure of their forelimbs. Indeed, it appears that the bats' ability to fly results from an accumulation of relatively small modifications, a series of gradual improvements, rather than through any single structural acquisition such as the feathers of birds. Thus, both heart and lung are typically mammalian, although relatively large. There is no paralleling of the special flow-through respiratory system of birds, despite similar high oxygen demands. On the other hand, there is no evidence that bats regularly fly at high altitudes like birds. This pattern of slow, steady refinement is also apparent at the physiological level too. Lung performance and the gas transportation properties of bat blood are likewise better than in other mammals, but not dramatically. There is, however, no question of the importance of these accumulated improvements, for it is possible to correlate their degree of refinement with life-style and flight performance from species to species.

The flight of bats, as we have seen, has evolved in parallel with their sonar system. Human hearing is limited to a band of frequencies between 20 Hertz and 20,000 Hertz, but bats are able to to hear ultrasonic frequencies up to about 200 kiloHertz. This, although previously suspected, was not demonstrated until 1938. Since then, it has been found that bats emit a wide range of calls, each associated with particular types of flight and behavior. Thus, European noctule bats (*Ayctalus noctula*) produce different sound patterns while migrating, cruising, searching for food, capturing food and when operating in close company with other bats.

Except for one genus of flying foxes, who use a unique and primitive system of tongue clicks to navigate out of the caves in which they roost, echolocation in bats is confined to the Microchiroptera, who use their powerful, bony larynx to generate the sound. Although many bats emit calls through their open mouth, others possess elaborate noses, using the accessory pleats and folds to modify and focus the sounds projected through their nostrils. So refined is the system that bats not only navigate with ease through confined spaces, accurately avoiding obstructions by millimeters, but are also able to use it to capture flying insects. This has led to an evolutionary arms race between bats and their intended prey, as each endeavors to outstrip the other. Noctuid moths, in particular, have proved adept at avoiding predation. Some species are able to tune in on the bat's sonar frequency. Changes in the signal reveal when the bat has located a potential victim, identified its flight path and velocity and begun closing in for the kill. Like a speeding driver alerted to the presence of police radar, the moth has only a brief moment to decide on evasive action. Some moth species close their wings to the body and drop like a stone as their nemesis approaches, so confounding the bat's predictions about its victim's flight path. Others, more technologically sophisticated, jam the bat's sonar transmissions by generating sounds of the same frequency. This either confuses the bat about the moth's location or imparts false information about its edibility. In either event, evolution advances another notch.

INSECTS AND FLIGHT

The first insects of which we have knowledge appeared in the fossil record during the Devonian era some 350 million years ago, and were flightless

Mayflies live as flying adults for only a few hours after spending many months as aquatic nymphs. Many individuals emerge at once, easily satiating predators and so ensuring that a significant number of these slow, tasty fliers survive to mate and lay eggs.

relatives of silverfish. The earliest winged insects are found in slightly younger deposits, about 300 million years old, and are quite clearly dragonflies. As pioneers in a new environment and free from competition, some of these early dragonflies grew to an enormous size. No insects since have come close to their 70-centimeter wingspan.

Since then the insects have flourished as no other group of animals. Fully three quarters of all life forms, and probably more, are insects. Despite this enormous proliferation of species, which has enabled insects to establish themselves in almost every type of nonmarine habitat, the range of structural diversification is not manifestly greater than that of other classes in the animal kingdom. Most insects are immediately recognizable as such. The success of insects may be attributed to the evolution of a wide variety of morphological, physiological and behavioral innovations, but none has been as profound and far-reaching as those associated with the power of flight.

Insect flight is one of the most challenging, tantalizing and frustrating areas of bioengineering. Unlike the flight of birds, in which a knowledge of conventional aerodynamics combined with intuition and sound experimental technique can usually provide solutions, the study of insect flight is a journey into parts unknown. Because of their small size, precise measurements of aerodynamic forces, velocity, pressure and the like are unbelievably difficult. The way in which the subject is restrained for study, the effects of anesthesia, the interference of the measuring devices themselves, these and other factors all conspire to make the data collected more than a little suspect. Add to this the influence of small Reynolds numbers and non-steady-state aerodynamic principles that are poorly understood, and it is not surprising that our knowledge of insect flight lags well behind our knowledge of bird flight.

This dragonfly has just shed its nymphal skin and emerged as an adult insect. Before it can fly the wings must swell to their full size and harden.

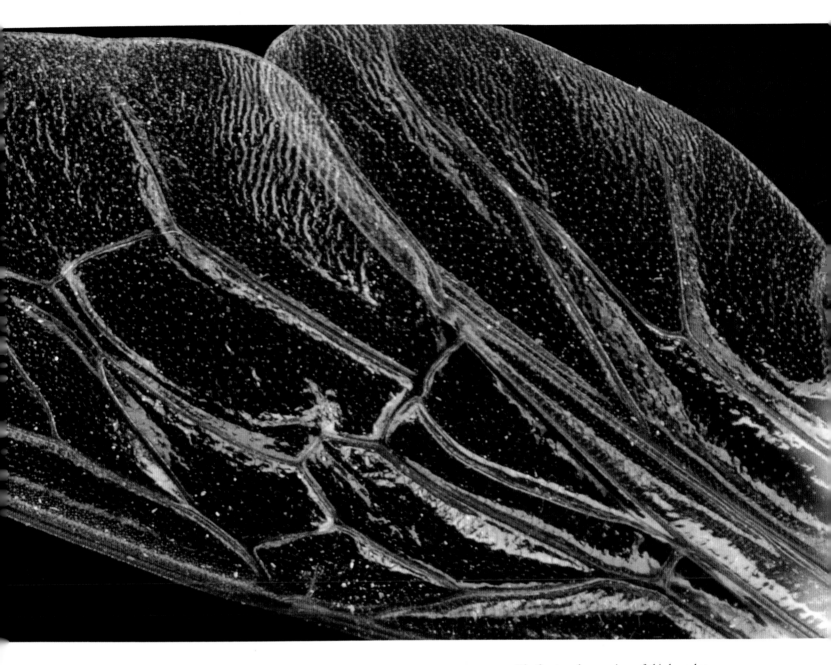

Insect Wings

Because the fossil record gives us only occasional glimpses at the course of evolution we really have no clear idea of how insect wings first arose. Consequently the subject has provoked long, intense and highly imaginative speculation. The general consensus today is that wings probably first evolved as heat-exchange surfaces projecting from the thorax. They would not have to protrude very far before starting to provide lift as an enhancement to jumping among stems and branches. We can observe a comparable evolutionary sequence in the Australian jumping spider *Saitis volans,* which has flaps of tissue projecting from the top of its abdomen. At rest these flaps envelop the body but when extended during jumping they provide significant lift and extend substantially the distance over which the spider glides.

Except for the complex joint where they attach to the thorax, insect wings have a deceptively simple structure. Only two to six microns thick, they consist of very thin sheets of cuticle separated by a network of veins. These are usually hollow and often contain both blood and fine, air-filled tracheal tubes. When the immature insect molts for the last time to become

The front and rear wings of this honeybee (Apis) are linked together by a system of tiny hooks that allow them to function as a single unit.

an adult, the wings are inflated by blood pressure to assume their final size and shape, and allowed to harden and dry. The wing is a marvel of engineering design, combining strength, stiffness, flexibility and aerodynamic performance with minimum weight and use of materials. The venation, apart from providing taxonomists with a valuable set of characters for identification, helps prevent buckling of the membranous wings. However, the main structural strength is supplied by a series of concertinalike pleated folds that run the length of the wing, particularly just behind the leading edge (Fig. 22). These have the added advantage that if the wing is injured, for example through impact with a twig, the stresses are widely distributed and no part of the structure seriously damaged. In a short time the wing regains its shape and function.

Seen in cross-section, insect wings appear to be the antithesis of all we would expect of an aerodynamic structure. Corrugated and irregular, with sharp ridges and deep valleys, they are in practice quite different from the smooth, filmy, gossamer structures they appear to the naked eye. Nowhere is this departure from our conventional view of wings more striking than among some of the very small insects such as plume moths and fairy flies. Here the wing is found to resemble a feather duster, or bottle brush, with a narrow central supporting core bearing a mass of long projecting filaments.

Typically, insects possess two pairs of wings, although the true flies, or Diptera, have only one functional pair, the hind wings being reduced to little clublike structures called halteres, which are part of a gyroscopic attitude-control system. Although among primitive flying insects such as termites and dragonflies the wings function independently, in most other orders one finds various devices for coupling the fore and hind wings into a single functional unit. These take many forms and include slots, bristles, hooks and interlocking hairs. Most insect wings appear smooth and membranous, although caddis fly wings are clothed in fine hairs, and those of butterflies and moths are covered in scales. The forewings of beetles are heavily armored to form protective covers, or elytra, for the delicate, propulsive hind wings, and a similar, though less pronounced, thickening of the forewings is also found among the crickets and grasshoppers (Orthoptera) and true bugs (Hemiptera).

Under the scanning electron microscope we can observe a marvelous array of curious surface structures. Some are undoubtedly sensory in function, but what of the others? By analogy with the wings of birds and aircraft, we can be sure that they enhance the wing's performance in some particular way. However, our present limited understanding of insect aerodynamics gives only occasional and very tantalizing hints as to what these roles might be.

The Insect Thorax and Wing Movements

Understanding how insects fly necessitates some knowledge of the thorax, the second of the three main divisions of the insect body. Although the thorax clearly originated through tagmosis in response to a reduction in the number of walking limbs, today this exercises only a secondary influence. To a very large extent the thoracic structure of higher insects is dictated by the functional demands of the wings and their associated flight muscles.

The thorax is a complex three-dimensional jigsaw puzzle composed of toughened plates held together by thinner, more flexible membranes. These elements are assembled to form a semirigid box, with limited ability to change shape. The insertion of the wings is quite straightforward in the more primitive orders, such as dragonflies. Here the wings emerge from near the top of the thorax, and with antagonistic pairs of muscles attached to their bases. Acting about a fulcrum at the point of insertion, these four

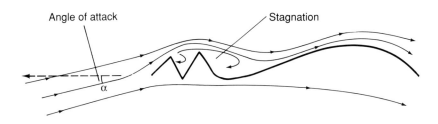

Fig. 22 Airflow over a Dragonfly Wing. Dragonfly wings acquire mechanical strength through corrugations just behind the leading edge. Although they lack the smooth contours of a bird or aircraft wing, dragonfly wings function similarly because they work at low Reynolds numbers. Still, viscous air trapped in the stagnation areas behind the corrugations effectively provides a smooth aerofoil surface. (After Rees.)

pairs of muscles provide power for both the up- and the downstroke of the wings. It is because the wings of dragonflies have independent, direct flight muscles, that they are able to beat out of phase with one another.

Among the more advanced, specialized fliers such as bees (Hymenoptera) and flies (Diptera), the arrangement of the wings and their connection to the thorax is markedly different. In these insects, the flight muscles act only indirectly on the wings. Thus, instead of attaching to the wing bases, as in dragonflies, the muscles that power the downbeat run between the top and bottom of the thorax, and the antagonistic pair that powers the upbeat runs horizontally through the thorax from front to back. Because lift is generated on both the up- and the downstroke, both sets of muscles are much the same size.

The key to understanding how the contractions of these two pairs of indirect flight muscles are converted into wing movements lies in the way in which the wings are articulated with the walls of the thorax. Inserted into the side walls of the thorax through a complex system of hinges and levers, the wings are oscillated up and down through slight distortions to

Caught at the moment of takeoff, this tropical chafer beetle has opened its armored front pair of wings to reveal the large, delicate hind wings that are used for propulsion. It is probable that some lift is generated by the rigid forewings, which mainly afford protection to the hind wings.

Insect flight muscles cannot work efficiently when cold. This shield bug is vibrating its wings before takeoff to increase the body temperature. Such behavior is frequently observed in bumblebees, which are well insulated and often fly early in the morning before other nectar feeders have warmed up.

the shape of the thorax. It takes only a small distortion of the thoracic wall to trigger a change in wing position, which means that the flight muscles have to alter only their tension rather than their length in order to function. Herein lies the secret of the extremely high wing-beat frequencies found in midges and other small insects.

It has long been thought that the wings of flies and other advanced insects are clicked up and down from one position to the other, rather like an electric light switch. By way of explanation, a variety of supposed mechanical benefits to be derived from such a click mechanism have been put forward. However, the click mechanism is not found in free-flying insects, and has now been shown to be an experimental artifact resulting from anesthesia and tethering, and the arguments explaining its possible advantages have been largely discredited.

Energy Needs

The power consumed by an insect beating its wings has two components. Part goes toward overcoming drag and generating lift, and part to overcoming the inertia of the wings themselves. Even though the wings are of very light construction, a significant amount of energy is required to accelerate and decelerate them several hundred times a second. In fact, more than half the energy consumed by a yellow jacket wasp in flight is absorbed by wing inertia. Further calculation reveals a curious paradox. The rate at which chemical energy appears to be converted into mechanical work falls far beyond the efficiency of any known muscular mechanism.

The solution to this enigma is to be found in the elastic nature of the thorax itself. If the thorax were constructed of rigid elements, the power generated by each muscle contraction would be dissipated at the beginning of the next return cycle. However, parts of the insect thorax are made of resilin, an extraordinary rubberlike protein. One of a number of such naturally occurring elastic substances, resilin is able to store energy generated by muscles and to release it as necessary at 97 percent efficiency. Thus, resilin is particularly important in insects such as flies, releasing energy into the system during the intermediate phases of the wing cycle. Resilin is also present in the knee joints of grasshoppers, and similar elastic energy-storage mechanisms are being found in a growing number of animal locomotory systems. The precise composition of these natural rubbers and their efficiency depends on the frequency at which they must function. Thus resilin, which is normally found in systems running well in excess of 100 Hertz, would not provide much benefit to a hummingbird, whose wings beat at around 50 Hertz.

Ironically, there is one special situation in which it is beneficial rather than detrimental to increase energy consumption. Flight muscles function most efficiently at temperatures above 30° C. Thus, in the early morning, most insects are immobilized by cold and have to wait until the heat of the sun warms them into activity. There are several exceptions to this general pattern, the most notable being bumblebees, which seem particularly well adapted to a life in cool, temperate climates. The bumblebee thorax is particularly well insulated and contains a number of heat conservation adaptations. By uncoupling their wings and contracting the flight muscles more slowly than usual, they are able to generate heat quickly in low ambient temperatures. Although bumblebees' high body and nest temperatures make them massive "gas guzzlers," their early start in the daily foraging race gives them ready access to rich supplies of nectar.

These yellow jackets (Vespula) *have discovered a convenient source of energy at a feeder set out to attract hummingbirds. Like bees, wasps link their wings to function as a single pair.*

Physiology of Insect Flight

It will be recalled from Chapter 1 that the flight muscles of insects are striated, their pattern of cross-banding reflecting the fine structure of sliding actin and myosin filaments. Because of this structure, flight muscles function only over a narrow range of lengths. Indeed, a fully extended fiber is only double the length of a fully contracted one, but the maximum force can only be exerted over a short distance, when the thin actin filaments have been drawn far enough into the thick myosin ones for all the cross-bridges to be activated. The speed of contraction — the speed at which the filaments slide past one another — is limited by the rate at which the cross-bridges can perform their cycle of attachment, and is inversely proportional to the force exerted.

This brings us to the question of wing beat frequency. In general, large animals operate at lower frequencies than small ones. Wildebeeste move their legs less rapidly than mice, and eagles flap their wings more slowly than sparrows. The same principle holds for insects, with dragonflies beating their wings at around 25 Hertz, bees at 250 Hertz and mosquitoes at 600 Hertz. Small midges fly at around 1000 Hertz. Quite apart from the metabolic demands of such operating frequencies, there is another major problem to be overcome.

Normally, flight muscle fibers contract in response to electrical impulses transmitted along the controlling nerve, each impulse producing a single twitch. Such muscles are called synchronous. It has long been thought that the upper frequency limit for the contraction of synchronous muscles is around 100 Hertz. Clearly the muscles of bees, wasps and mosquitoes must be different and possess special properties. Called asynchronous, or fibrillar, muscles, they respond to a single triggering impulse by initiating a self-perpetuating cycle of oscillation. The stiff elasticity of the thorax, combined with the tension of the muscles, form a tuned system in which the fibers vibrate in synchrony at high frequencies with little change in length. The rhythm is controlled and maintained by an occasional timing impulse and continues until an inhibitory signal is received.

It has always been assumed that asynchronous muscles arose in response to the need for high wing-beat frequencies, but recently this interpretation has been questioned because of the discovery of a growing number of synchronous muscles able to operate at high frequencies well above 100 Hertz. Indeed, the synchronous tymbal muscles of a cicada, which are responsible for generating its penetrating song, have been found to function at 550 Hertz. Thus, it appears that asynchronous muscles may have arisen not because of their high operating frequency but rather because they confer greater metabolic efficiency with reduced weight.

Fortunately for researchers, several large beetles and water bugs that have evolved from small, fast-flying ancestors still retain fibrillar muscle systems with big fibers. Even so, studying the physiology of such systems is far from easy.

Insect flight muscles are efficient converters of chemical energy, and as such require a substantial supply of oxygen. It will be recalled that birds, which face similar metabolic problems, have managed to improve the flow of air through their lungs. Because insects are so much smaller than birds, they have been able to reach a different solution. Instead of relying on hemoglobin circulating in a well-plumbed circulatory system to transport oxygen and carbon dioxide to and from distant respiratory surfaces, gas exchange occurs within the muscles themselves. This is done through a network of fine tubes radiating and branching throughout the body. In the more active fliers with greater metabolic demands, pumping movements of the abdomen combined with controlled openings of the spiracles creates a one-way flow-through pattern of air movements along the main tracheal trunks, resulting in more efficient gas exchange.

The Aerodynamics of Insect Flight

A clue to the mysteries of insect flight lies in their size, for they function at relatively low Reynolds numbers. A precise understanding of insect aerodynamics may never be possible, because small differences in size or velocity produce a confused mixture of viscous and inertial forces on different parts of the wing at different stages of the beat cycle. Moreover, it is becoming clear that, depending on the type of wing, different mechanisms are involved. Thus the flight of dragonflies is quite different from that of large beetles, butterflies or minute parasitic wasps. And what about the flight of lacewings like *Croce* (Neuroptera)? The hind wings of these ant lion relatives are drawn out into long streamers three or four times the length of the owner's body. Whether they simply provide stability or are used in some way as aerial paddles is unknown. Presumably, if they provided any outstanding benefits, such structures would be more frequent and widespread.

Because of their low Reynolds numbers, insect wings do not suffer the grave inertial consequences that would plague larger aerofoil surfaces with similar contours. The valleys and dead spaces that one might suppose would generate turbulence and drag become filled with slowly circulating bubbles of air. These effectively provide the wing with a functional skin of air over which lift forces can be generated (Fig. 22). By this means, the wing inertia can be kept as low as possible. Presumably, the same principle applies even more dramatically to the bottle-brush wings of plume moths and fairy flies, for whom the air is correspondingly more viscous because of their very small size. Although conventional aerodynamic forces undoubtedly play a significant role in insect flight, especially in large species, mysterious and little-understood transient effects are being increasingly implicated. These were first identified by Torkel Weis-Fogh, a Danish professor at Cambridge University in England. Working with a tiny parasitic wasp called *Encarsia,* Weis-Fogh discovered that lift in this creature was generated by a wholly new and unsuspected mechanism, which he came to name "clap-and-fling." *Encarsia,* like a hummingbird, holds its body vertical, vibrating its wings horizontally. It was noticed in experiments that maximum lift was generated just as the wings began to peel apart, having clapped together at the end of the upstroke. That is, lift appeared to be associated with the separation of the wings and not with their accelerating downbeat. From this observation, it was eventually discovered that the lift was being produced by the intermittent circulation of the bound vortex about the wing. It will be recalled from Chapter 2 that the bound vortex is an obscure and hitherto largely theoretical concept, whose existence was required to counter the starting vortex generated behind the trailing edge of a wing. We now know that the bound vortex can account for the lift forces generated by an aerofoil in flight.

Variations on Weis-Fogh's "clap-and-fling" mechanism are constantly being discovered as more and more insects are subjected to detailed scrutiny. For example, in the green lacewing *Chrysopa,* the wings meet on both the up- and downstroke, generating lift twice in each cycle. What makes *Chrysopa* even more remarkable is that the two pairs of wings are wholly out of phase with one another.

The flight patterns of butterflies are distinctive and their massive wings seem unnecessarily large and cumbersome for the weight they have to bear. It is known that the patterns and colors of butterfly wings play important roles in temperatue regulation and mimicry, but it is doubtful that such distinctive structures could have arisen solely to satisfy such nonaerodynamic needs. There still remains much to learn, but it is now known that the cabbage white butterfly *(Pieris brassicae)* holds its body horizontal and claps the wings together above its head. The air movements that begin as the wing tips start to peel apart continue to grow as the wings are flung

The large head of this damselfly (Ischnura) is part of an inertial guidance system. Note the relatively massive size of the thorax, which houses the antagonistic pairs of flight muscles.

downward and end up as a large doughnut-shaped vortex beneath the insect. The resultant of the forces that generate these vortices is the lift that keeps the butterfly airborne. This observation suggests that the vortex-ring theory of flight, developed to explain the aerodynamic forces generated by birds, applies equally well to insects, where it is enhanced by the special properties of the clap-and-fling mechanism and its derivatives.

In still air, the clap-and-fling pattern of wing beats has been found to enhance lift by about 25 percent. This being so, one wonders, why has it not been adopted by all flying insects — and birds as well? Presumably the answer must lie in the fact that some penalty is incurred in the use of clap-and-fling flight. Perhaps rapid forward flight is compromised through the excessive drag generated by large, flat wings.

A clearer understanding of the unsteady aerodynamics of insect flight has come from the study of dragonflies. These rapacious aerial pirates have flown virtually unaltered for over 200 million years, and hence may be regarded as a well-proven and successful design. Careful analysis of the airflow over the two pairs of wings, which in dragonflies beat half a cycle out of phase with each other, reveals that vortices are formed by the forewings as their speed and angle of attack are constantly altered. These vortices accelerate air quite precisely over the hind wings, greatly enhancing their lift.

Control and Stability

In contrast to human pilots, who lack the innate ability to recognize sudden changes in speed and attitude, birds, bats and insects, on the other hand, have evolved the necessary sensory and control ability as part of their total flight package. This is particularly true of insects, whose evolutionary investment in sophisticated flight control systems has paid off in enhanced agility and maneuverability. In this respect, there appears to be a clear tendency for the number of wings to be functionally or actually reduced to a single operating unit. Thus, in bees and wasps a row of hooks on the leading edge of the hind wing, the frenulum, engage in a groove along the trailing edge of the forewings so they function as a single pair.

As previously noted, the Diptera, or true flies, which are the most specialized insect fliers, have their hind wings reduced to halteres. These drumsticklike structures provide the flies with a three-dimensional counterpart to an aircraft's gyro compass. Hinged at their point of insertion on the thorax behind the wings, the knobbed ends of the halteres rotate in a regular pattern as the insect flies. Any changes in speed or attitude will modify slightly the path described by the weighted ends of the halteres. In steady flight, stretch receptors at the bases of the halteres monitor their movements, barraging the fly with a steady pattern of nerve impulses. Changes in the path of the halteres, resulting from pitching, for example, will alter the timing of signals from the stretch receptors. Analysis of these differences within the fly's nervous system automatically generates signals to the appropriate wing muscles. Slight changes in the relative timing and twisting of the wings bring about the required correction, although the precise details are not yet understood. Some flies are even able to disengage one wing entirely and fold it back along the body while in flight. Insects, while inherently unstable, have to make such corrections many times a second, conferring on themselves superb maneuverability. A fly about to be swatted is able to perceive the impending disaster and take the necessary evasive action in five milliseconds or less. In contrast, the human eye would not even have detected any movement in that short a time span, let alone initiated any response.

A different form of inertial guidance system is found in dragonflies, whose ability to capture prey on the wing depends on sharp vision and

great maneuverability. The dragonfly's massive compound eyes make the head large and heavy. Connected to the body by a thin, highly flexible neck, the inertia of the head relative to the body provides the dragonfly with information similar to that which the halteres provide to the fly. In dragonflies the actual sense organs feeding this control system are located between the head and thorax in the region of the neck.

True flies (Diptera) have only a single pair of functional wings. The hind pair are modified into special knobbed sensory organs called halteres. These can be are clearly seen on this crane fly (Tipula).

CONCLUSION

Our journey of exploration into the remote byways of the animal kingdom is now completed. Although the reader may feel overwhelmed by the diversity of life forms encountered and the variety of methods whereby they move, in truth we have barely scratched the surface. Interdisciplinary studies are pushing back our horizons at an ever-increasing rate, and scarcely a week goes by without some significant advance in biomechanics being announced in the scientific literature. We rightly marvel at the immensity of our knowledge and understanding. Future generations, provided any animal life survives for study in the wild, can look forward to even more wonderful revelations about the outcome of natural selection. Nowhere, I venture to predict, will the lode prove richer than among the insects.

APPENDIX:
THE CAST OF
CHARACTERS

Bacteria

The bacteria are not animals, but belong to a separate, very ancient kingdom, the Monera, which also includes blue-green algae. In general, bacterial cells are far smaller than those of protists and animals, and exhibit many structural differences indicative of distinctive biochemical activities. They receive mention here because some species are motile and employ a unique form of flagellar locomotion.

Protists or Protozoa

Now afforded the status of a separate kingdom, along with the slime molds, diatoms and yellow-green algae, many protists have traditionally been regarded as part of the animal kingdom. Protists were, at one time, thought of as single-celled (unicellular) organisms. However, the individual protist is functionally comparable to a whole multicellular animal, rather than to just one of its component cells. Structural elements within the protist body, known as organelles, fulfill roles equivalent to the organs of the Metazoa, or multicelled animals. It is, therefore, better to regard protists as acellular rather than unicellular.

The cilia and flagella used by protists form an important component of our story. Flagellate species belong to the Mastigophora and ciliate species to the Ciliophora or Ciliata. The Sarcodina, which include the foraminiferans, radiolarians and amoebas, move by cytoplasmic streaming, a generalized flowing of the cell contents that is also known as amoeboid motion.

Sponges (Phylum Porifera)

Sponges possess a very simple level of organization with minimal coordination between their component cells. They figure in here only because they contain large numbers of special cells, called choanocytes, that possess flagella similar to those found in protists. These are used to create water currents for feeding and respiration. In addition, sponges have a mobile larval stage known as a planula, which propels itself by means of cilia.

Coelenterates (Phylum Cnidaria)

These are multicellular animals whose body is arranged on an essentially radial plan. There are three classes, the Hydrozoa, the Scyphozoa (true jellyfish) and the Anthozoa (sea anemones). The Hydrozoa can possess both mobile jellyfishlike medusae and sessile polyps in their complex life cycles. They include both the classic polyp *Hydra*, beloved of school biology classes, and elaborate colonial forms such as the Portuguese man-of-war, which belongs to the order Siphonophora.

Comb Jellies (Phylum Ctenophora)

The ctenophores somewhat resemble the true coelenterate jellyfish in their form and organization. However, they lack the stinging cells of coelenterates, replacing them instead with special adhesive cells. In addition, the comb jellies are characterized by special swimming organs called comb plates, which are formed from uniquely fused cilia.

Flatworms (Phylum Platyhelminthes)

We are concerned here with only the free-living turbellarian flatworms, ignoring the related flukes and tapeworms, whose parasitic life-styles largely preclude the need for locomotion. Flatworms are the most primitive of the bilaterally symmetrical Metazoa, and possess a solid body with no coelom or other body cavity. The surface layer of cells, the epidermis, bears cilia on the underside of the body. Beneath the epidermis is a thin fibrous layer, the basement membrane, and below that are two layers of muscle, the outer one circular (running around the body) and the inner one longitudinal. The hydrostatic skeleton of flatworms consists of gelatinous parenchyma cells.

The turbellarian flatworms range in size from primitive Acoela no larger than ciliated protozoa to large tricladids several centimeters in length. Freshwater tricladids are often called planarians, and there are also a few species that have become terrestrial.

Ribbon, Proboscis or Nemertean Worms (Phylum Nemertina)

Members of this small phylum of marine worms are larger and somewhat more highly organized than the flatworms, from which they are apparently derived. Most ribbon worms are enormously elongated, the European bootlace worm, *Lineus longissimus,* growing to over 30 meters in length although only a few millimeters in diameter. Shallow coastal waters are their typical habitat, where they may be found tanged in a Gordian knot beneath stones. However some species have become terrestrial, and superficially resemble flatworms.

The principal difference in structure that concerns us here is the presence in nemertines of an eversible proboscis, which lies above the mouth and gut within a fluid-filled chamber extending through the front third of the body. Resembling the finger of a glove turned inside out, the proboscis and its associated hydraulic system provides a limited form of fluid skeleton analogous to the coelom or hemocoel of higher invertebrates. This is in addition to the ancestral hydrostatic skeleton provided by parenchyma cells like those of flatworms.

Rotifers (Phylum Rotifera)

Rotifers are among the smallest of the Metazoa, and most are less than 0.5 millimeter long. The great majority live in fresh water, but a number are quite salt-tolerant and are found in marine and estuarine habitats. Although some can crawl like flatworms or loop like leeches, most swim actively by means of a well-developed corona of cilia.

Roundworms (Phylum Nematoda)

The nematodes are a highly successful group with numerous species, both free-living and parasitic. Some are no larger than ciliate protozoans, but others, such as the human intestinal parasite *Ascaris,* grow to almost half a meter in length. The relevant features of their morphology are the possession of a tough cuticle, longitudinal muscles but no circular ones and a fluid-filled body cavity, the pseudocoel. This functions like the coelom, but is not homologous with it, being differently derived embryologically. The cuticle is highly elastic, and contains two sets of opposing fibers wrapped helically around the body.

Proboscis Worms (Phylum Priapulida)

The priapulids are a small group of rather obscure, unpretentious burrowing marine worms, whose body contains a coelom. This is a particular type of body cavity typical of more highly organized animals, and is widely used as a hydrostatic skeleton. Both longitudinal and circular muscles are present beneath the tough outer cuticle.

The distinguishing characteristic of the priapulids is a massive spiny proboscis, or introvert, bearing the mouth at its tip. The introvert is used for burrowing in mud and engulfing prey animals that crawl near the burrow entrance.

True or Segmented Worms (Phylum Annelida)

Almost 9000 annelid species are known, the largest reaching more than 3 meters in length. Annelids are mentioned quite often in this book, not only because of their varied methods of locomotion, but also because they play a key role in discussions about the origin of arthropods. The annelids are divided into three main classes: Polychaetes or bristleworms; Oligochaetes, which include earthworms; and the Hirudinea or leeches. Lying beneath a thin external cuticle are layers of circular and longitudinal muscle, and between the body wall and the tubular gut there runs a fluid-filled cavity, the coelom, extending the whole length of the animal. The most striking feature of all annelids is their metameric segmentation — that is to say, the repetition of virtually identical elements along the length of the whole animal. Thus each segment includes a section of the gut, of the coelom, of the longitudinal blood vessels and of the ventral nerve cord, as well its own musculature. In many species the portion of coelom in each segment is effectively isolated from its neighbors, providing the hydrostatic skeleton with a high degree of localized muscular control. However, for some lifestyles this not an advantage, and the intersegmental septa that divide the coelom may be secondarily lost.

The polychaetes, which are virtually all marine, generally have a pair of lateral projections on each body segment called parapodia. These parapodia bear numerous spines and can be moved by muscles running back into the body. In contrast to the polychaetes, oligochaetes bear no parapodia and have few spines (chaetae) along their body. In addition to the familiar earthworms, the Oligochaeta include numerous small aquatic species, found mainly in fresh water.

Leeches (Hirudinea) are the most specialized annelids. The majority live in fresh water, but some are marine and others terrestrial. All are either predaceous or ectoparasitic, and bear adhesive suckers at each end. Segmentation is greatly reduced, the body is somewhat flattened and there are no parapodia or chaetae. Our interest in the leeches stems from the two special methods of locomotion they have evolved, and which immediately distinguishes them from the other annelids. Both are associated with the replacement of the coelom as a hydrostatic skeleton by specialized botryoidal cells, which impart a necessary stiffness that would otherwise be lacking.

Mollusks (Phylum Mollusca)

The mollusks are a large and diverse group comprising chitons, gastropods, bivalves and cephalopods, together with a few other more obscure animals. Although all mollusks share a fundamental body plan, evolution has effectively concealed it with a wealth of subsequent, far-reaching modifications and adaptations.

Although the mollusks possess a coelom, it is largely suppressed and reduced in size, and it performs no skeletal function. This role has been taken over by a secondary body cavity, the hemocoel, which consists of an ill-defined spongy network of sinuses and small vesicles through which the blood permeates. The ancestral segmented coelomate body plan is found today only in *Neopilina*, a relict deep-water species from the Caribbean. In others, the segmented body plan has become obscured beneath a host of later evolutionary changes.

Perhaps the two most conspicuous features of the phylum are a shell and a powerful muscular foot, although both may be reduced or lost in some forms. Apart from some specially adapted air-breathing gastropods, the mollusks respire through gills. These lie in a special chamber, the mantle cavity, which has also been subjected to many evolutionary modifications in different groups.

The chitons (Polyplacophora) are widespread seashore animals protected above by heavily armored overlapping plates. They adhere to the rocks on which they live by means of a powerful muscular foot.

The gastropods are an enormously successful group with over 105,000 living species, the largest growing to about 1 meter in length. They are divided into three groups: the prosobranchs, the opisthobranchs and the pulmonates. The prosobranchs include most of the typical shell-bearing marine gastropods such as the whelks, limpets, conchs and cowries. The opisthobranchs include the sea slugs and sea butterflies, and are aquatic forms, most of whom have lost their shell. The pulmonates include the typical garden slugs and snails. All are air-breathing, even though some have subsequently adopted a life in fresh water and, except for the slugs, bear a shell into which they can withdraw.

The body of bivalve mollusks is protected by hinged pairs of shells (called valves), which can be clamped tightly shut when necessary. More than 20,000 living species have been described, the largest growing to 1.35 meters in length and with a weight exceeding 200 kilograms. They possess a protrusible foot, homologous with that in gastropods, and they breathe through gills lying within a space called the mantle cavity. In clams and oysters the mantle cavity opens directly to the surrounding water, but burrowing species often have a long siphon through which to draw in and expel the water that carries both oxygen and food particles.

The most specialized mollusks are the cephalopods, a distinct class that includes the nautilus, squid, cuttlefish and octopus. All can swim actively by jet propulsion, and modifications to their mantle cavity and its associated structures have been adapted to this end. Although only about 650 species of cephalopod are alive today—including the largest of all invertebrates, squid, measuring 16 meters in length—they were at one time far more abundant. More than 7500 fossil species have been described, the oldest dating back to the Cambrian.

Peanut Worms (Phylum Sipunculoidea)

Sipunculids belong to a small phylum of unsegmented burrowing and boring marine worms found mainly in shallow water. In many ways they resemble the priapulids, with which they used to be classified. Thus they are coelomate, with both longitudinal and circular muscles. Like the priapulids they also possess a proboscis or introvert with the mouth at the tip, but the sipunculid introvert is long and thin, and bears a crown of ciliated tentacles called a lophophore.

Arthropods

All animals with jointed limbs used to be united into one enormous phylum called the Arthropoda that included insects, spiders, crustaceans, myriapods and various lesser-known forms. However, it is clear that the arthropods do not represent a single evolutionary line, united by common descent, but rather sprang independently and at different times from various wormlike ancestors. The "arthropod" features that were thought to unite them have actually arisen quite independently as convergent mechanical solutions to separate but similar evolutionary demands.

Phylum Chelicerata

This assemblage includes the arachnids (spiders, harvestmen, scorpions, false scorpions, solpugids, etc.), and the merostomes, a largely extinct group represented today only by the horseshoe crabs. A third group of uncertain affinities, the sea spiders (Pycnogonida), are also placed in this phylum. The arachnids have the body divided into two units or tagmata, the cephalothorax and abdomen, and typically possess four pairs of walking legs.

Phylum Crustacea

A large group, mostly aquatic, in which the limbs are branched (biramous) and extremely diverse in form. The crustacean body is conspicuously segmented behind the carapace, most segments bearing specialized paired appendages. Crustaceans also possess two pairs of antennae, which are sometimes modified for swimming in addition to their primary sensory function.

The main crustacean taxa are the Class Branchiopoda, which includes the water fleas (Cladocera); the Class Copepoda; the Class Branchiura, or fish lice; the Class Cirripedia, or barnacles; and the Class Malacostraca. The Malacostraca are a large group that includes sand hoppers (amphipods), wood lice or sow bugs (isopods) and shrimps, crabs and lobsters (decapods).

Phylum Uniramia

This is a newly recognized taxon that includes three subphyla, the velvet worms (Onychophora), the hexapods and the myriapods. The hexapods include the true insects (Class Pterygota) the springtails (Class Collembola) and silverfish (Class Thysanura). The myriapods include the centipedes (Chilopoda) and millipedes (Diplopoda) as well as two less-familiar groups, the classes Symphyla and Pauropoda, whose members lack common names.

The true or pterygote insects comprise some 26 different orders, divided into two groups depending on their manner of growth and development. The most advanced have immature larval and pupal stages in their life histories that are quite different in form from the adult stage. These include the two-winged flies (Diptera), beetles (Coleoptera), fleas (Siphonaptera), butterflies and moths (Lepidoptera) and the bees, ants and wasps (Hymenoptera). The remainder pass through an immature nymphal stage that is similar to the adult stage, except in the development of the wings. This group includes the dragonflies (Odonata), grasshoppers (Orthoptera), termites (Isoptera), and the true bugs (Hemiptera).

The members of this phylum all have similar unbranched limbs, similar mandibulate mouthparts and similar locomotory gaits.

The Acorn Worms (Phylum Hemichordata)

Hemichordates comprise a small group of marine worms that excavate burrows in mud. At one time they were thought to be allied to the chordates because they have gills, but this relationship is now discounted. They are not very active animals, but nevertheless have a highly modified hydrostatic skeleton. The coelomic cavity has become largely occluded by connective tissue and muscle cells developing in the epithelial lining. This strange coelomic musculature has taken over many of the functions of the original body-wall musculature, but the advantages resulting from this substantial reorganization remain obscure.

Echinoderms (Phylum Echinodermata)

These are wholly marine animals whose bodies are arranged about a pentamerous (five-rayed) symmetry. They are protected by calcareous plates and move by means of protrusible, hydraulic tube-feet. The echinoderms include the starfish (Asteroidea), the brittle stars (Ophiuroidea), urchins (Echinoidea) and sea cucumbers (Holothuroidea).

Chordates (Phylum Chordata)

In additional to the familiar vertebrate groups, fish, amphibia, reptiles, birds and mammals, the chordates include some rather less well-known animals. Among these are the sea squirts or ascidians, and some rather obscure relatives. *Amphioxus,* the lancelet, is a simple fishlike creature that is placed in its own separate subphylum.

The earliest vertebrates were the jawless fishes (Agnatha), of which the lamprey and hagfish are the sole living survivors. With the demise of the Agnatha during the Devonian (345–405 million years ago), a second group of armored fish, the placoderms, arose. They are important for having given rise to the Amphibia, and are the direct ancestors of lungfish and the coelacanth. Then came the cartilaginous fish or elasmobranchs, which today are represented by the sharks, skates and rays. The great majority of living fish possess bony skeletons and are known as teleosts.

GLOSSARY

Acoelomate Lacking a coelom. The lower wormlike invertebrate phyla.

Aerofoil A surface such as a wing designed to generate high lift and low drag.

Alveolus The smallest air chambers within the lung.

Amoeboid Resembling an amoeba in having no fixed form and moving by means of pseudopodia.

Bernoulli's Theorem Relates wind velocity and pressure distribution over an aerofoil surface. Pressure is least where the velocity is greatest.

Biomass The weight of living organisms in a given habitat.

Biramous Forked or divided into two branches.

Bivalve Mollusks such as clams and scallops that possess a hinged shell.

Botryoidal Resembling a bunch of grapes. A term used to describe specialized cells in the body of leeches having a skeletal function.

Cambrian Most ancient of the Palaeozoic rocks, dating back about 600 million years.

Cephalothorax The anterior of the two body divisions of a spider.

Chaetae Stout spines, particularly those projecting from the bodies of annelid worms.

Coelom The main body cavity of many animals, lined with epithelial cells and of great evolutionary significance.

Commensal A mutually beneficial association between two organisms.

Coxa The segment of an arthropod leg closest to the body.

Cryptic An animal that conceals itself and gains protection by resembling dead leaves and other inert objects.

Devonian Paleozoic rocks about 400 million years old that contain the first insect and amphibian fossils.

Epithelial A special form of cells associated with surfaces.

Fibrillar A special type of muscle powering the wings of higher insects in which a single nerve impulse triggers a cycle of contractions.

Hemocoel The major secondary body cavity of mollusks and arthropods, filled with blood and with no connections to the outside.

Heterotrophic Feeding on dead or living remains of other organisms.

Inchworm Caterpillars of moths of the family Geometridae that move with a characteristic looping movement.

Isometric A form of muscle contraction that does not involve a change of length.

Laminar Flow over a surface in which streamlines do not vary with time and in which there is no turbulence.

Metamerism   The repetition of similar segments along the body of an animal.

Myriapod   A general term referring to centipedes, millipedes and their relatives.

Parenchyma   A soft spongy tissue in which cells are separated by a gelatinous matrix of fluid-filled spaces.

Peristalsis   Waves of contraction passing along the length of the body.

Permian   Most recent of the Paleozoic rocks, dating back about 280 million years. The period in which amphibians declined and reptiles expanded in abundance.

Phylum   One of the thirty-odd major divisions of the animal kingdom, containing animals of similar levels of organization.

Planarians   A general term for certain freshwater and terrestrial turbellarian flatworms.

Plankton   Organisms that drift near the surface at the mercy of ocean currents.

Polychaetes   Members of one of the three main divisions of the annelid worms, characterized by having many chaetae or spines.

Precambrian   Rocks of great antiquity, predating the Paleozoic era and hence more than 600 million years old. Rocks of this age contain few fossils.

Protist   A general term for any member of the kingdom Protozoa.

Pulmonates   Members of a group of air-breathing gastropod mollusks, some of which live under water.

Quadruped   Any animal that walks on four legs.

Scaling   The study of changes that result between organisms because of differences in size or scale.

Sinusoidal   Alternating in a series of regular S-shaped waves.

Sternite   A sclerotised plate on the undersurface of an arthropod.

Streamline   An imaginary line in a moving fluid connecting points of equal velocity.

Tagmosis   Division of the body into a series of functional rather than anatomical segments.

Tergite   A sclerotised plate on the upper surface of an arthropod.

Test   The hard shell of urchins and other echinoderms.

Toroid   Ring-shaped, like a doughnut, with a hole in the center.

Trochanter   The segment of an arthropod limb attached to the coxa.

Undulate   Move with a gentle wavelike motion.

Vortex   Tornado-shaped rotation in a fluid.

PHOTO CREDITS

SELECTED READINGS

The literature on animal locomotion is proliferating rapidly. Most contributions are to be found in specialized and highly technical scientific journals. It is neither appropriate nor possible to give a detailed listing here. However, it should be noted that the *Journal of Experimental Biology* is particularly rich in papers dealing with insect flight.

The following books are important sources and provide much basic information. They will also serve as the starting point for a deeper search of the scientific literature on locomotion:

Alexander, R. McNeill. *Locomotion of Animals.* Cambridge, MA: Blackwell, 1982.

_____ *Animal Mechanics.* Cambridge, MA: Blackwell, 1982.

Calder, W. A. *Size, Function and Life Histories.* Cambridge, MA: Harvard University Press, 1984.

Elder, H. Y., and Trueman, E. R. *Aspects of Animal Locomotion.* New York: Cambridge University Press, 1980.

Gray, J. *Animal Locomotion.* London: Weidenfeld & Nicholson, 1968.

McMahon, T. A., and Bonner, J. T. *On Size and Life.* New York: Scientific American Library, 1983.

Pedley, T. J. (ed.) *Scale Effects in Animal Locomotion.* New York: Academic Press, 1977.

Schmidt-Nielsen, K. *Scaling.* New York: Cambridge University Press, 1984.

Thompson, D. W., and Bonner, J. T. *On Growth and Form.* New York: Cambridge University Press, 1961.

Trueman, E. R. *Locomotion of Soft-Bodied Animals.* New York: Elsevier, 1975.

Wainright, S. A., Biggs, W. D., Currey, J. D., and Gosline, J. M. *Mechanical Design in Organisms.* Princeton, NJ: Princeton University Press, 1976.

INDEX

Italic numbers indicate illustrations

beat frequency of 176
in insects 171–174
theory of 31–33
Wood louse 112, 188
Woodpeckers 124, *132*
Work 25
Worm(s)
  acorn (hemichordata) *80, 96,*
  189
  bootlace (*Lineus longissimus*)
  185
  burrowing 11
hydrostatic skeleton of 13
nemertean (nemertine) 88,
  185
peacock (*Sabella*) *100*
peanut (sipunculids) 187
polychaete 47
proboscis (priapulids) *90, 91,*
  95, 185–186
quill (*Hyalinoecia*) *94*
rag (*Nereis*) *48,* 102–103,
  103
ribbon 13, 89, *90,* 185
segmented (*Annelids*) 9, 13,
  96–97, 186
velvet (Onychophora) 105–
  106, *108,* 188
See also Flatworms, Round-
  worms
Wright brothers 27, 161

X

Xerus 136

Y

Yellow jackets (*Vespula*) 175

Z

Zebras *137*
Zirphaea 98
Z-line 5